Panayotis C. Yannopoulos
Aristeidis A. Bloutsos

Air Pollution in the University of Patras Campus, Greece

Panayotis C. Yannopoulos
Aristeidis A. Bloutsos

Air Pollution in the University of Patras Campus, Greece

Monitoring details and data evaluation for the period 2012-2013

LAP LAMBERT Academic Publishing

Impressum / Imprint

Bibliografische Information der Deutschen Nationalbibliothek: Die Deutsche Nationalbibliothek verzeichnet diese Publikation in der Deutschen Nationalbibliografie; detaillierte bibliografische Daten sind im Internet über http://dnb.d-nb.de abrufbar.

Alle in diesem Buch genannten Marken und Produktnamen unterliegen warenzeichen-, marken- oder patentrechtlichem Schutz bzw. sind Warenzeichen oder eingetragene Warenzeichen der jeweiligen Inhaber. Die Wiedergabe von Marken, Produktnamen, Gebrauchsnamen, Handelsnamen, Warenbezeichnungen u.s.w. in diesem Werk berechtigt auch ohne besondere Kennzeichnung nicht zu der Annahme, dass solche Namen im Sinne der Warenzeichen- und Markenschutzgesetzgebung als frei zu betrachten wären und daher von jedermann benutzt werden dürften.

Bibliographic information published by the Deutsche Nationalbibliothek: The Deutsche Nationalbibliothek lists this publication in the Deutsche Nationalbibliografie; detailed bibliographic data are available in the Internet at http://dnb.d-nb.de.

Any brand names and product names mentioned in this book are subject to trademark, brand or patent protection and are trademarks or registered trademarks of their respective holders. The use of brand names, product names, common names, trade names, product descriptions etc. even without a particular marking in this work is in no way to be construed to mean that such names may be regarded as unrestricted in respect of trademark and brand protection legislation and could thus be used by anyone.

Coverbild / Cover image: www.ingimage.com

Verlag / Publisher:
LAP LAMBERT Academic Publishing
ist ein Imprint der / is a trademark of
OmniScriptum GmbH & Co. KG
Heinrich-Böcking-Str. 6-8, 66121 Saarbrücken, Deutschland / Germany
Email: info@lap-publishing.com

Herstellung: siehe letzte Seite /
Printed at: see last page
ISBN: 978-3-659-64074-2

CONTENTS

1

1. INTRODUCTION

1.1 The University of Patras Campus

The University of Patras was founded in the city of Patras in 11th November of 1964 by the Presidential Decree No. 4425. The first operational academic year was 1966-1967. In 1968, an area of 2.66 km^2, 12 km NNE of the city center, was chosen for the construction of the University Campus at the S of Rion Village and on the foot of Panachaicon Mountain. The University Campus includes 24 Departments with a large number of sectors. Specifically, the University consists of five Schools: The School of Natural Sciences, The School of Engineering, the School of Health Sciences and the School of Humanities and Social Scientists (www.upatras.gr). The population is approximately 30,000 including students, faculty members and other personnel. There are about 750 Professors and Lecturers, 450 persons belonging to administrative staff and approximately 26,000 undergraduate and 2,000 postgraduate students. Nowadays, there are about 230 building blocks in various sizes and for different purposes, having a total area of 257,000 m^2. These buildings consist of more than 200 laboratories, a Central Library, a Church, a Science and Technology Museum, a Botanic Museum, a Zoological Museum, a Museum of Education, a Conference and Culture Center, a bank, a post-office, coffee shops and restaurants, dormitories, a swimming pool, institutes of research, computing centers and other facilities. At the N boundary of University Campus with the settlement of Kato Kastritsi there are the University Sports Center and School buildings, including a nursery school, primary and secondary schools. In the Campus area there is also the Regional University Teaching Hospital, which functions both as the major regional medical center and as a teaching facility for the Faculty of Medicine. There have been extensive works of primary and secondary network infrastructure, sports facilities, extensive plantings and other of significant importance.

The Environmental Engineering Laboratory (EEL) of the Department of the Civil Engineering has run several air pollution monitoring programs in the major Patras area. Since April of the year 2012, EEL has started operating an air pollution monitoring Station, which is installed within the University of Patras Campus (UPC), as shown in Fig. 1 and 2, under the responsibility of the first author.

The present book presents and evaluates the treated data of the EEL air pollution monitoring Station measured during April 2012 to December 2013 and compares them to other measurements in urban, suburban and rural areas of Greece and other Mediterranean cities.

3

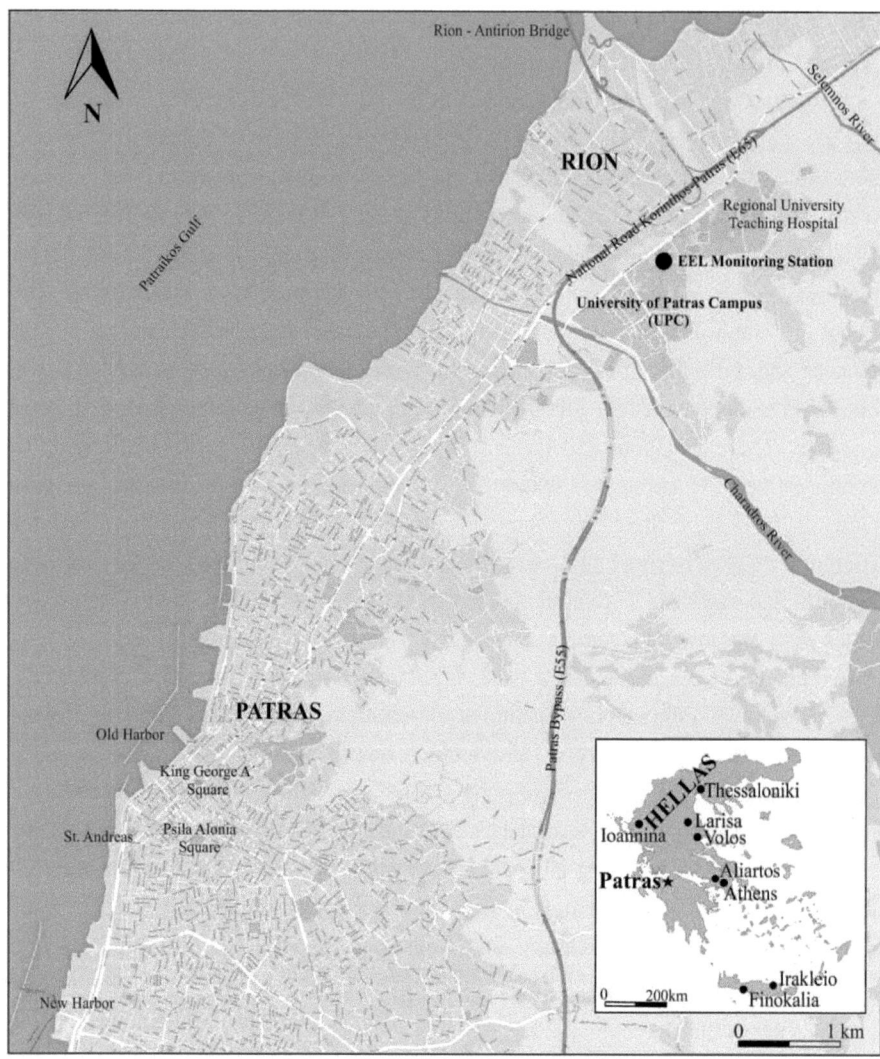

Fig. 1 Map of major Patras area. (Source: http://www.ploigos.gr)

1.2 Air Pollution Monitoring Station

The position of the EEL Monitoring Station of UPC is shown in Fig. 2. The EEL Station is located at the western parking lot of the Building of the Department of Civil Engineering (geographical longitude 21°47′22′′, geographical latitude

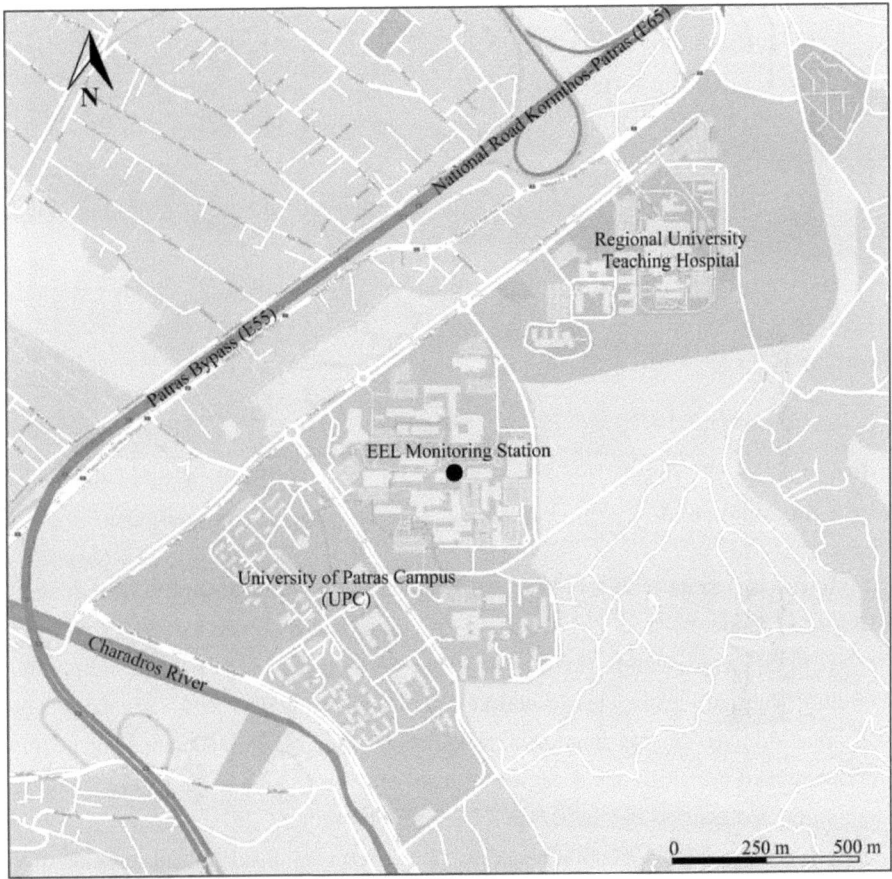

Fig. 2 Map of the Campus of the University of Patras.
(Source: http://www.ploigos.gr)

38°17′22′′ and 60.60 m altitude above sea level). At the area of station's installation the inclination is 4-5% toward NW. Apart from asphalt-covered streets, the major area consists of natural soil with low vegetation, bushes, and sporadic trees, mainly pine and olive trees. The Station is settled at a distance more than 15 m W from the 3-storey building of the Civil Engineering Department, while all other buildings are even further away. At a radius of 0.7 km N, approximately, the old National Road and the new National Road Korinthos – Patras (E65) pass. At the same direction at a distance 2.2 km the local ferry port of Rion – Antirrion is. The Patras By-Pass (E55) is 1.2 km SE away from the Station. The University Hospital of Patras is located 1.5

(a)   **(b)**

Fig. 3 Air Pollution Monitoring Station: (a) External view; (b) internal view

km NE it. Also, more than 2 km toward NE, there is a limited number of industrial activities of moderate size. The Station is free from nearby objects of any kind from the NE to SE wind sectors (i.e. for an angle of at least 247.5°). The Campus air pollution originates from classic sources of a suburban-rural area, augmented by emissions due to central heating during winter and additional emissions from aforementioned activities and a cement factory operating 2-3 km NE of the Campus. The Station is classified as "Suburban – background station".

The housing of the EEL Station consists of thermally insulated aluminum, with dimensions 212×144×284 cm (length×width×height) and is settled on the sidewalk pavement (Fig. 3a). A meteorological station that records continuously the temperature, humidity, barometric pressure, wind speed and wind direction has been installed on the EEL Station roof. Inside the building the temperature is maintained in the range (20-25°C). The Station is equipped with automatic analyzers (Fig. 3b) in continuous operational mode recording data of Particulate Matter (PM_{10}, $PM_{2.5}$ and PM_1), Ozone (O_3), Nitrogen Oxides (NO, NO_2 and NO_x), Carbon Monoxide (CO) and Sulfur Dioxide (SO_2). The most important characteristics regarding analyzers and measurements are given in Table 1.

The calibration of the installed automatic analyzers is made using gas calibrator equipment (Sabio Instruments, Inc. Model 4010 –Gas Dilution Calibrator and Sabio Instruments, Inc. Model 1001-Zero Air Source, Fig. 4a). The gas calibrator equipment uses known concentration standards dense mixtures of air pollutants in nitrogen stored in bottles of standard concentration to produce diluted air pollutant mixtures of known concentration. The produced diluted gas is drained to the analyzer input that is set for calibration. For the calibration procedure, the bottles shown in Fig. 4b, which contain air pollutants with the following standard concentrations, are used:

1. Standard concentration of 10 ppm carbon monoxide (CO)
2. Standard concentration of 1000 ppm carbon monoxide (CO)
3. Standard concentration of 30 ppm sulfur dioxide (SO_2)
4. Standard concentration of 30 ppm nitrogen monoxide (NO)
5. Standard concentration of 500 ppm nitrogen monoxide (NO)
6. Standard concentration of 2 ppm nitrogen monoxide (NO)

Table 1. Analyzer and measurement characteristics

Pollutant	Analyzer	Measurement Concept	Start Date
Particulate Matter (PM_{10}, $PM_{2.5}$ και PM_1)	Grimm Aerosol Technik, Mobile Dust Monitor, Environ Check 180	90° scattering light measurement	07/04/2012 for PM_{10} & $PM_{2.5}$ 07/09/2012 for PM_1
Sulfur Dioxide (SO_2)	Dasibi 4108-UV Fluorescence	UV Fluorescence	03/06/2013
Carbon Monoxide (CO)	Monitor Labs, 9830B	IR Absorption	11/03/2013
Ozone (O_3)	Monitor Labs, 9810B	UV Absorption	12/03/2013
Nitrogen Oxides (NO, NO_2 και NO_x)	Horiba, APNA-370	Chemiluminescence	19/02/2013

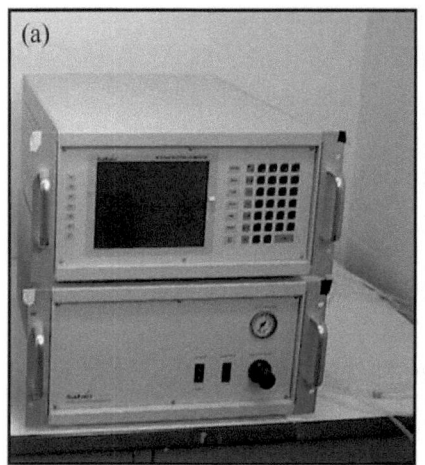

Fig. 4 Gas calibrator equipment: (a) Gas Dilution Calibrator; (b) bottles containing air pollutant gases of standard concentrations.

2. AIR POLLUTANTS

Human activities are the main factor that is responsible for the majority of air pollutants that are emitted in the atmosphere. According to their origin, air pollutants are classified to (a) primary air pollutants (PM, SO_2, CO, NO), that are emitted directly from pollutant sources and (b) secondary air pollutants (PM, O_3, NO_2), that are formed in the atmosphere by chemical reactions or physical processes. Epidemiological studies, many of which are reviewed by World Health Organization (WHO), European Environmental Agency (EEA) and US Environmental Protection Agency (US-EPA) (WHO 2006, 2008; 2013a, b; US-EPA 2009; EEA 2013), show that air pollution causes adverse effects on human health. Also, air pollution affects ecosystems by damaging vegetation, downgrading freshwaters and soil quality and causing adverse effects on animals similar to humans. Finally, air pollution is related to climate change in several ways, although they are distinct problems.

2.1 Carbon Monoxide (CO)

Carbon monoxide is a colorless, tasteless, odorless and enough poisonous gas. It is produced by the oxidation of organic matter in an oxygen-deficient environment (incomplete combustion). It is emitted from car exhausts, chimneys of industrial plants and heating sources that use wood as combustible material. The usual CO levels are higher during cold periods, because the low temperatures make combustion less complete and cause inversions that trap pollutants close to the ground. The CO concentrations follow the diurnal traffic patterns, showing elevated values at urban areas especially during rush hours where traffic occurs. It is noted that, since the usage of catalytic converters, CO emissions have significantly reduced. Carbon monoxide does not appear to have a negative impact on plants and materials, at least with its current levels in the atmosphere. At higher concentrations CO causes serious problems in the human aerobic metabolism. Prolonged exposure in environments with carbon monoxide, even at low concentrations, causes adverse health effects ranging from dizziness, fatigue and headaches and even up to death. As CO contributes to the formation of greenhouse gases, such as CO_2 and ozone, it affects the climate change (WHO 2006; US-EPA 2009; EEA 2013).

2.2 Sulfur Dioxide (SO_2)

Sulfur dioxide is a colorless, non-flammable and non-explosive gas, having an irritated odor at high concentrations and it is readily soluble in water. Although its

9

natural sources, such as volcanoes, contribute to the formation of its physical levels, the most important sources of SO_2 emissions locally are anthropogenic. It is produced from the combustion of solid or liquid fuels that contain sulfur at domestic heating, stationary power plants and motor vehicles. The effects on human health depend on the concentration levels and exposure time. They are mainly noticed by the people with respiratory and lung problems. Also mortality and hospital admissions for the cardiac disease are increased during the days with high SO_2 levels. Sulfur dioxide contributes to acid precipitation (acid rain), which affects negatively vegetation, causing discoloration and damages to the plants foliage. In addition, acid rain reduces the atmospheric visibility and increases the acidity of river and lake waters. Regarding materials, the presence of SO_2 is associated with corrosion of steel and other metals, while decays building's coating materials and particularly those containing carbonates, such as marble, limestone and mortar (WHO 2006; US-EPA 2009; EEA 2013).

2.3 Nitrogen Oxides (NO_x)

Nitrogen oxides are inorganic compounds, including compounds of the form NO_x, such as nitrogen monoxide (NO), nitrogen dioxide (NO_2), and other rarer compounds, as nitrogen protoxide (N_2O), dinitrogen pentoxide (N_2O_5) and some others. For air pollution monitoring, NO and NO_2 is of primary importance. While NO is a non-irritating, colorless and odorless gas, NO_2 is irritating having a reddish brown color and a characteristic pungent odor. On global scale, emissions of NO_x are due to human activities rather than natural sources. In nature, NO is produced by both lightning and bacteria found in natural soil and plants through the process of denitrification. The production of NO_x is made by the chemical reaction between nitrogen and oxygen occurring at high temperature, as it happens during the combustion of volatile organic compounds into the burners. The main human activity responsible for this production is the burning of fossil fuels at stationary sources (i.e. heating, power generation) and in mobile sources (i.e. internal combustion engines of vehicles and ships). The major part of NO_x emissions are NO emissions, which are rapidly oxidized in the atmosphere to other NO_x gases. Only a small part of NO_x emissions is directly emitted as NO_2 for most combustion sources except of diesel vehicles, because their exhaust after-treatment systems increase the direct NO_2 emissions. In general, NO_2 is mainly formed by oxidation of the reactive NO gas. Oxidation of NO by atmospheric oxidants, such as O_3, occurs rapidly, even at the low levels of reactants. On the ecosystem, NO_x cause serious adverse effects, as the retardation of plant organisms' growth. Regarding human health, NO_x are associated with adverse effects. The liver, lung, spleen and blood are affected by NO_2. It also aggravates lung diseases, which may lead to respiratory symptoms and increased susceptibility to respiratory infection of the respiratory system (WHO 2006; US-EPA 2009; EEA 2013).

2.4 Particulate Matter (PM_{10}, $PM_{2.5}$ and PM_1)

Particulate matter (PM) is a mixture of airborne particles that are consist of solids and liquid droplets, having a wide range in size and chemical composition and are suspended for extended period of time in air. Until nowadays, there is a considerable importance to particles with aerodynamic diameter less than 10 μm (PM_{10}) known as "inhalable particles" capable to deposit in the upper respiratory tract and those having aerodynamic diameter less than 2.5 μm ($PM_{2.5}$) known as "fine particles". PM is either directly emitted (i.e. from chimneys) as primary particles or it forms (secondary particles) in the atmosphere from the oxidation and transformation of primary gaseous emissions such as SO_2, NO_x, NH_3 and VOCs (volatile organic compounds, a class of chemical compounds whose molecules contain carbon), known as precursor gases. The presence of airborne PM is due to natural (naturally suspended dust, Saharan dust, sea salt, pollen, volcanic ash etc.) and anthropogenic (central heating, industry, road traffic, wildfires etc.) sources. Especially in urban areas, important local sources include road traffic (vehicle exhausts, emissions from abrasion and re-suspension processes and road dust re-suspension, Gehrig *et al.* 2004; Ketzel *et al.* 2007), burn of wood, fuel or coal for domestic heating and small industries.

Airborne PM has adverse effects on human health and environment. Epidemiological studies have shown a strong association between PM and mortality. The health effects of PM are caused after the exposure to them. The inhalation and penetration of PM into the lungs and blood stream lead to the appearance or aggravation of cardiovascular, neural and respiratory diseases. These effects depend on the particle size of airborne material. The finer the size of airborne, the deeper penetrates into the respiratory system and the correlation between mortality and particulate matters becomes more important. Investigators have also linked PM_{10} air pollution to morbidity and hospital admissions. It is remarkable that the risk of various outcomes has been shown to increase with exposure; so it is difficult to suggest a threshold below which no adverse effects would be anticipated. The environmental impact may be assessed by the temporary occurrences of particulate matter that affect visibility, climate and vegetation. In addition, building materials do not remain unaffected from their exposure to particulate pollution. The effect is the soiling of buildings and even a corrosive effect on them depending on PM's composition (WHO 2006; US-EPA 2009; EEA 2013).

2.5 Ozone (O_3)

Ozone is a colorless gas and the main component of photochemical smog. It is clearly a secondary air pollutant. Unlike to primary pollutants, which are emitted directly into the air, ground-level (tropospheric) O_3 is formed from complex photochemical reactions following emissions of precursor gases, such as NO_x, CO, methane (CH_4) and non-methane volatile organic compounds (NMVOCs) in the presence of sunlight. It is destroyed by reacting with NO_2 and is deposited to the ground. Concentration levels of O_3 are increased during summer season. Due to its highly reactive chemical properties, O_3 is harmful to human health, vegetation and materials. High levels of O_3 may cause respiratory health problems, including decreased lung function, aggravation of asthma and other lung diseases. Several studies have shown a correlation with mortality. Elevated levels of O_3 can also damage vegetation leading to reduced agricultural crop yields, decreased forest and reduced biodiversity. In addition to the above effects, O_3 is an efficient greenhouse gas contributing to the warming of the atmosphere (WHO 2006, 2008; US-EPA 2009; EEA 2013).

2.6 Air Quality Standards and Guidelines for selected air pollutants

An important consequence of the Industrial Revolution was the reduction of air quality due to human activities. During 19th century a rapid increase of industry occurred first in Europe and extended quickly at N of the United States of America (USA). This increase followed by a decrease of the air quality at those areas. The degradation of air quality was initially realized by the adverse effects due to the presence of black smoke emitted at industrial cities by coal combustion. Until 50's all the efforts were made to reduce smoke emissions in the atmosphere. But, two serious events that occurred in Los Angeles in 1943 and London in 1952 showed that a new form of air pollution, known as "smog", was responsible for increasing mortality and hospital admissions. Since then, a substantial increase has occurred to research and monitoring, both in Europe and the United States (Bachman J., 2007). These efforts led to the first US National Ambient Air Quality Data (NAAQS) by US-EPA in 1971 and the first EC Directives by European Union (EU) in 1970 and 1972. First NAAQS where stricter than EC Directives as they were introduced limit values for CO, O_3, total suspended particulate (TSP), NO_2 and SO_2 instead of EC Directives, that settled measures against gases emitted from engines of motor vehicles (http://www.air-quality.org.uk, November 2014). Since those years, several reviews have been made to NAAQS by US-EPA and new Directives were introduced by the EEA of EU, setting limits and target values of air pollutants, as a result of the intense research that took place in the scientific community. Furthermore, WHO introduced Guidelines, the latest in 2006 (WHO, 2006) of air pollutants for the protection of human health and the environment. Nowadays, the existing NAAQS are published by US-EPA's website (http://www.epa.gov November 2014) including the airborne pollutants of CO, lead (Pb), NO_2, O_3, $PM_{2.5}$, PM_{10} and SO_2. The EU Directive 2004/107/EC, relating to arsenic (As), cadmium (Cd), mercury (Hg), nickel (Ni) and polycyclic aromatic hydrocarbons (PAHs) in ambient air, and the latest EU Directive 2008/50/EC on ambient air quality and cleaner air for Europe, which settles air quality target and limit values for the air pollutants of CO, Pb, NO_2, NO_x, O_3, $PM_{2.5}$, PM_{10}, SO_2 and benzene (C_6H_6), are currently among the strictest acts of legislation worldwide (Gemmer, M & Xiao, B, 2013). The WHO Guidelines (WHO, 2006) include particulate matter ($PM_{2.5}$ and PM_{10}), O_3, NO_2, SO_2 and CO.

The above standards and guidelines, of selected air pollutants that were measured by the EEL monitoring station at the UPC, are summarized in Table 2.

Table 2. Air Quality Standards and Guidelines for selected air pollutants monitored at the University of Patras Campus (UPC)

Air pollutant	Averaging period	EEA (Directive 2008/50/EC) Concentration	EEA Comment	US – EPA (http://www.epa.gov) Concentration	US – EPA Comment	WHO (2006) Concentration
$PM_{2.5}$	Calendar Year	25 µg m⁻³	To be met by January 2015	12 µg m⁻³	Annual mean, averaged over 3 years for Primary Standards	10 µg m⁻³
	Calendar Year	20 µg m⁻³	To be met by January 2020			25 µg m⁻³
	24-h Mean			35 µg m⁻³	98th percentile, averaged over 3 years for Primary and Secondary Standards	
PM_{10}	24-h Mean	50 µg m⁻³	Not to be exceeded on more than 35 days per year	150 µg m⁻³	Not to be exceeded on more than once per year on average over 3 years for Primary and Secondary Standards	50 µg m⁻³
	Calendar Year	40 µg m⁻³				20 µg m⁻³
O_3	8-h Mean	120 µg m⁻³*	Not to be exceeded on more than 25 days per year averaged over three years	150 µg m⁻³ (0.075 ppm)	Annual 4th highest daily maximum 8-h concentration, averaged over 3 years for Primary and Secondary Standards	100 µg m⁻³*
NO_2	1-h Mean	200 µg m⁻³	Not to be exceeded on more than 18 hours per year	200 µg m⁻³ (100 ppb)	98th percentile, averaged over 3 years for Primary Standards	200 µg m⁻³
	Calendar Year	40 µg m⁻³		100 µg m⁻³ (53 ppb)	Annual Mean for Primary and Secondary Standards	40 µg m⁻³
$NO_x=NO+NO_2$	Calendar Year	30 µg m⁻³	Vegetation			

Table 2 (continued)

Air pollutant	Averaging period	EEA (Directive 2008/50/EC) Concentration	EEA Comment	US – EPA (http://www.epa.gov) Concentration	US – EPA Comment	WHO (2006) Concentration
	1-h Mean	350 µg m^{-3}	Not to be exceeded on more than 24 hours per year	200 µg m^{-3} (75 ppb)	99th percentile of 1-h daily maximum concentrations, averaged over 3 years for Primary Standards	
	24-h Mean	125 µg m^{-3}	Not to be exceeded on more than 3 days per year			20 µg m^{-3}
SO$_2$	Calendar Year	20 µg m^{-3}	Vegetation			
	Winter**					
	Mean 10 min			1300 µg m^{-3} (0.5 ppm)	Not to be exceeded on more than once per year for Secondary Standards	500 µg m^{-3}
CO	1 h Mean			40 mg m^{-3} (35 ppm)	Not to be exceeded on more than once per year for Primary Standards	30 mg m^{-3}
	8-h Mean	10 mg m^{-3}*		10 mg m^{-3} (9 ppm)		10 mg m^{-3}*

* Maximum Daily 8-h Mean

** 1 October – 31 March

15

3. PRESENTATION AND DISCUSSION OF MEASUREMENTS

The elaborated concentration measurements of airborne PM (PM$_{10}$, PM$_{2.5}$ and PM$_1$), O$_3$, NO$_x$ (NO+NO$_2$) and CO measured at the University of Patras Campus (UPC) are presented in diagrams. The recorded SO$_2$ concentrations were quite small, at background levels. Therefore, it is decided not to consider in the present work. It is noted that the elaborated PM, O$_3$, CO και SO$_2$ concentrations measured at UPC are based to the 5 min values recorded by the automatic analyzers. The elaboration of NO$_x$ values is based on 8-h records of the analyzer. The time that is indicated on charts is related to the current time, i.e. including changes between winter and summer seasons.

3.1 Variation of daily mean pollutant concentrations

The variation of daily mean concntrations of PM$_{10}$ during the years 2012 and 2013 is presented in Fig. 5 and 6, respectively. Considering that PM$_{10}$ is a primary air pollutant, we can see that the daily values are lower than the limit value of 50 μg m^{-3} (Directive 2008/50/EC). This value exceeded 4 times during the measurement period in the year 2012 and 7 times during 2013. These exceedances are due to natural

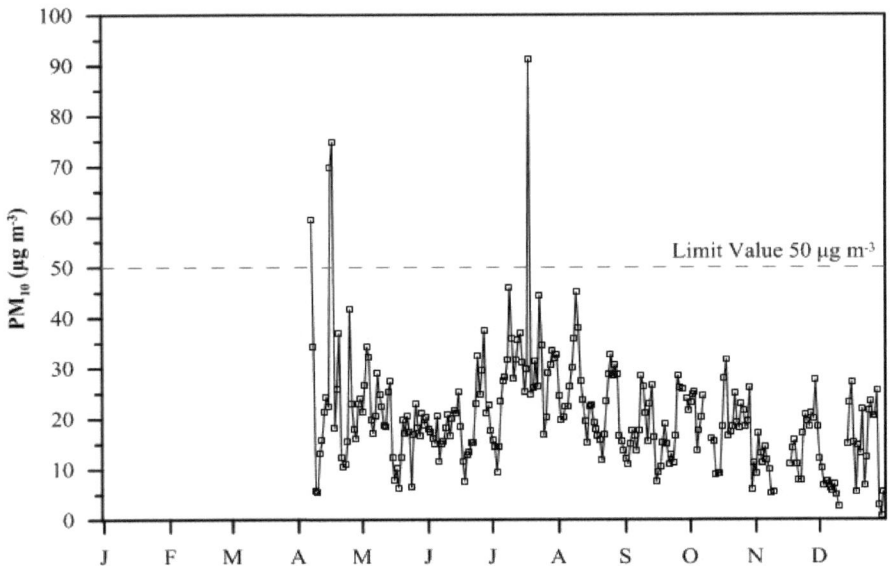

Fig. 5 Daily mean PM$_{10}$ concentrations during 2012, measured at the UPC site

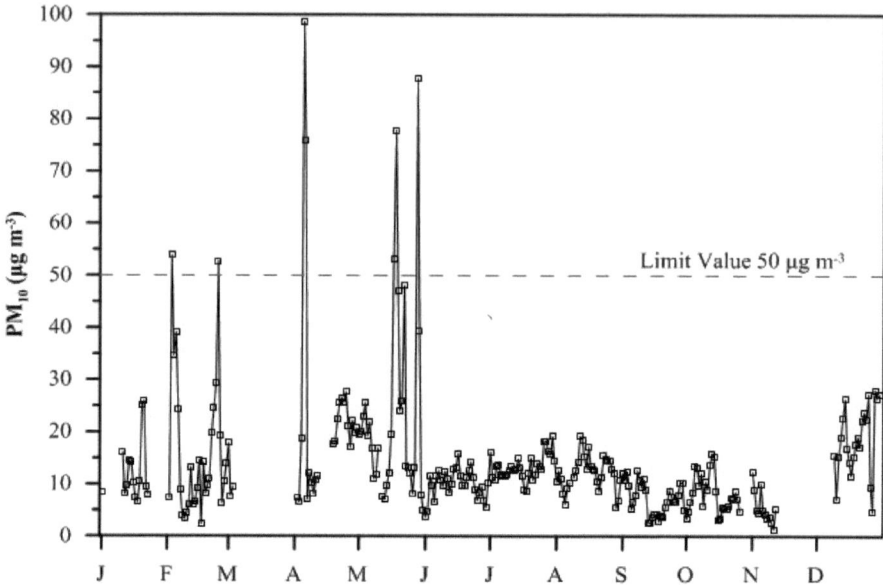

Fig. 6 Daily mean PM₁₀ concentrations during 2013, measured at the UPC site

activities, such as the trasported Sahara's dust from Africa and the fire events took place in the wider Patras area during 18 to 25 of July 2012.

Figures 7 and 8 show the variation of daily mean concntrations of $PM_{2.5}$ during the years 2012 and 2013, respectively, measured at the UPC site. The daily values are lower than the limit value of 25 $\mu g \ m^{-3}$ (Directive 2008/50/EC). The aforementioned value was exceeded only 5 times during the measurement period in the year 2012 and 9 times during 2013. These exceedances, as in the case of PM_{10}, are due to the same activities.

The variation of daily mean concntrations of PM_1 during the measurement period in the year 2012 and in the year 2013, measured at the UPC site, is presented at Fig. 9 and 10, respectively. The concentration values are less than 25 $\mu g \ m^{-3}$ during the measurement period, but neither the USA nor EU have established by now a specific limit or target value for PM_1.

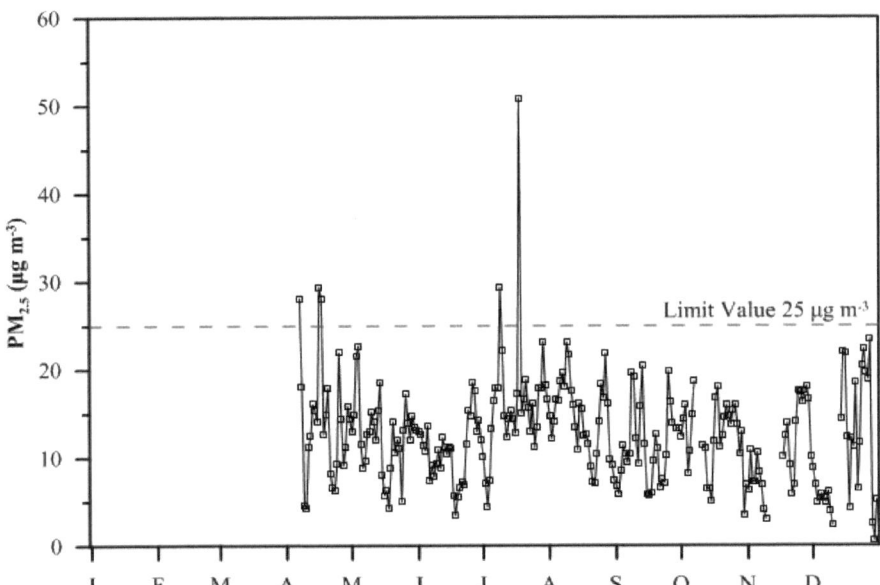

Fig. 7 Daily mean PM$_{2.5}$ concentrations during 2012, measured at the UPC site

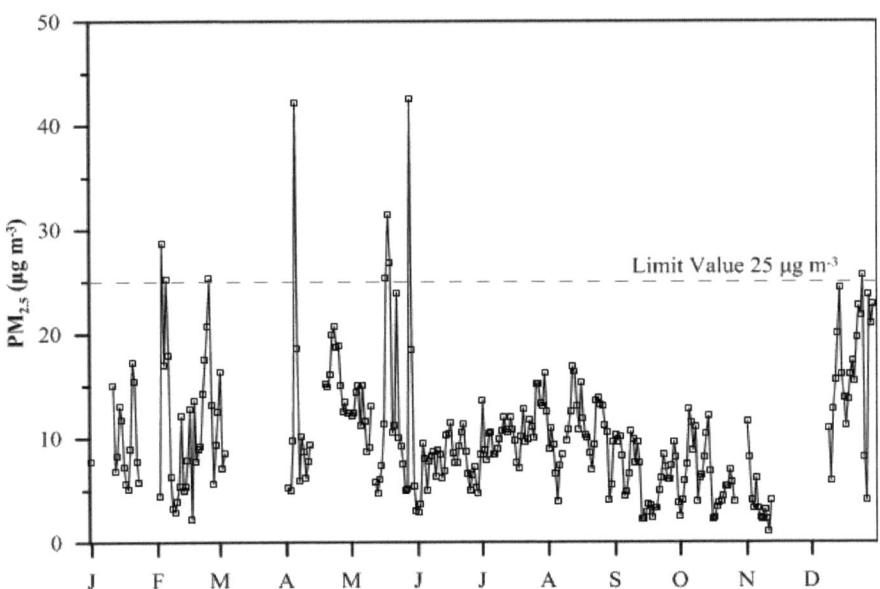

Fig. 8 Daily mean PM$_{2.5}$ concentrations during 2013, measured at the UPC site

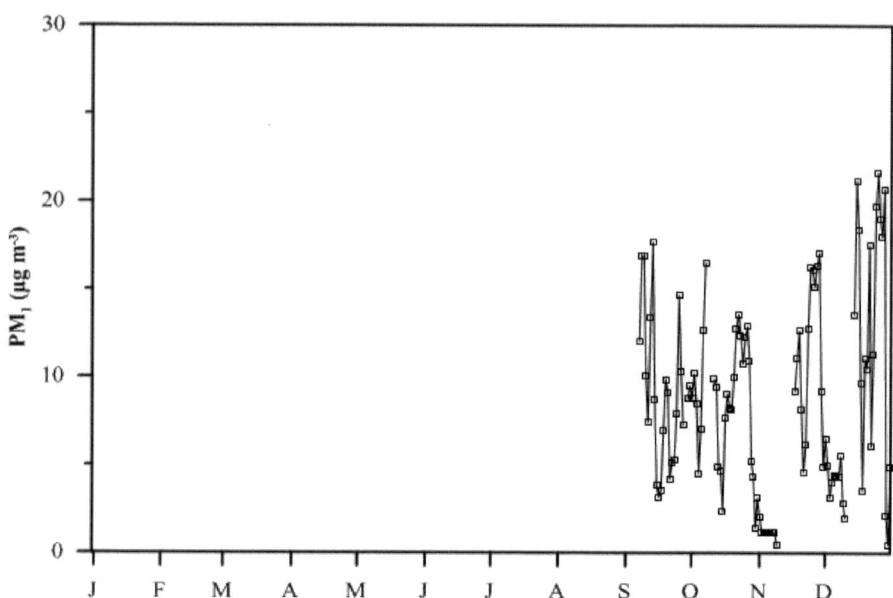

Fig. 9 Daily mean PM$_1$ concentrations during 2012, measured at the UPC site

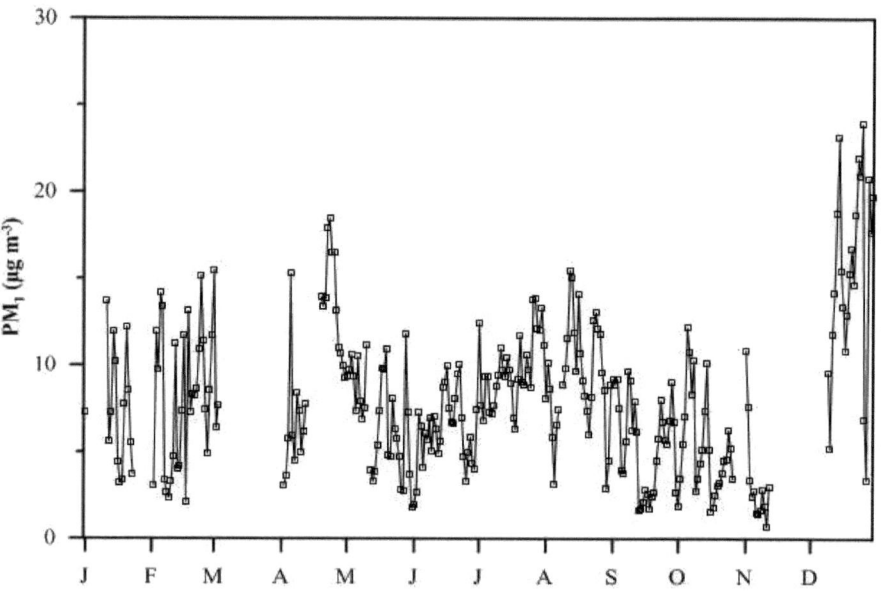

Fig. 10 Daily mean PM$_1$ concentrations during 2013, measured at the UPC site

Fig. 11 Satelite images from MODIS on several dates showing Sahara's dust events over Patras area (Source: https://earthdata.nasa.gov/labs/worldview/)

An example of evident presence of Sahara's dust events over wider Patras area during the measurement period is shown in Fig. 11. Images of Moderate Resolution Imaging Spectroradiometer (MODIS), captured by NASA's Aqua Satelite and provided through NASA's website (https://earthdata.nasa.gov/labs/worldview/, show that on 09/07/2012, 19/05/2013 and 29/05/2013 there were plumes of dust originating from N. Africa and covering the wider Patras area. These events explain the PM peaks that occurred during those days. Sahara's dust events are also responsible for some other PM peaks, but the related satelite images are neither available nor clear due to cloudiness.

The variation of daily mean O_3 concentrations, measured at the UPC site, during the measurement period in the year 2013 is presented in Fig. 12. The daily mean values were calculated as the mean value of the related 8-h average values recorded. During the measurement period in the year 2013, 148 exceedances of the limit value of 120 $\mu g\ m^{-3}$ were noticed, a number much greater than permitted. Note that the 120-$\mu g\ m^{-3}$ limit must not be exceeded more than 25 times per calendar year averaged over three years (Directive 2008/50/EC).

The variations of daily mean concentrations of NO, NO_2 and NO_x during the measurement period in the year 2013, measured at the UPC site, are presented in Fig. 13, 15 and 17, respectively. These values were calculated as the mean value of the related 8-h average values recorded. The variation of daily maximum concentrations of NO, NO_2 and NO_x for the year 2013 are presented in Fig. 14, 16 and 18, respectively. The daily mean values were calculated as the mean value of the related maximum 8-h average values recorded. The mean and maximum daily values of NO_x were found quite low and, since neither limit nor target values have been established for mean daily values, they indicate that NO_x may not be a matter for UPC.

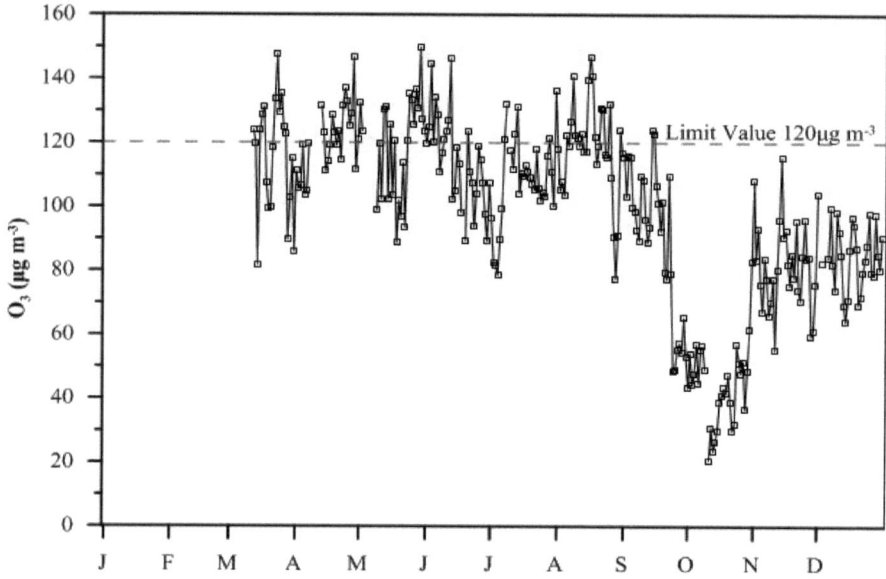

Fig. 12 Daily mean O_3 concentrations during 2013, measured at the UPC site

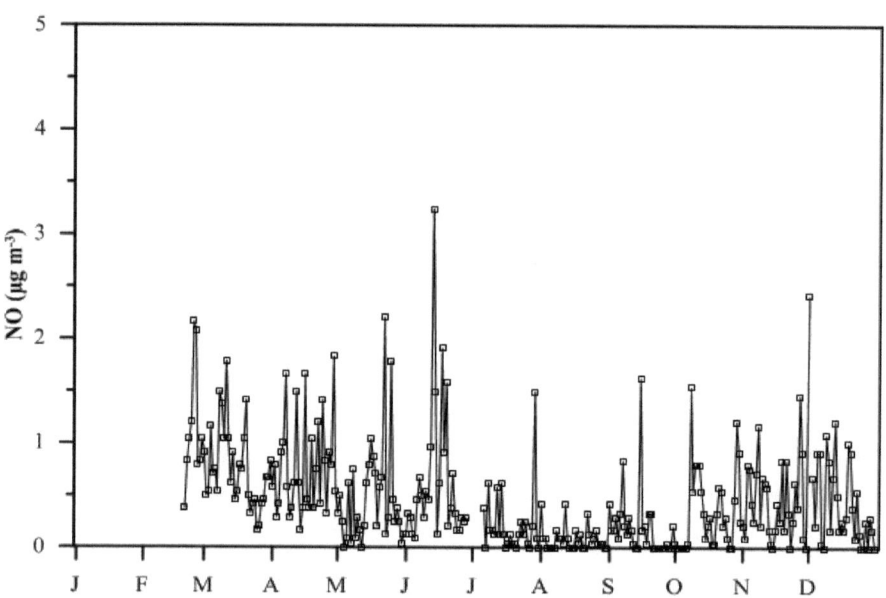

Fig. 13 Daily mean NO concentrations during 2013 (derived from mean 8-h values)

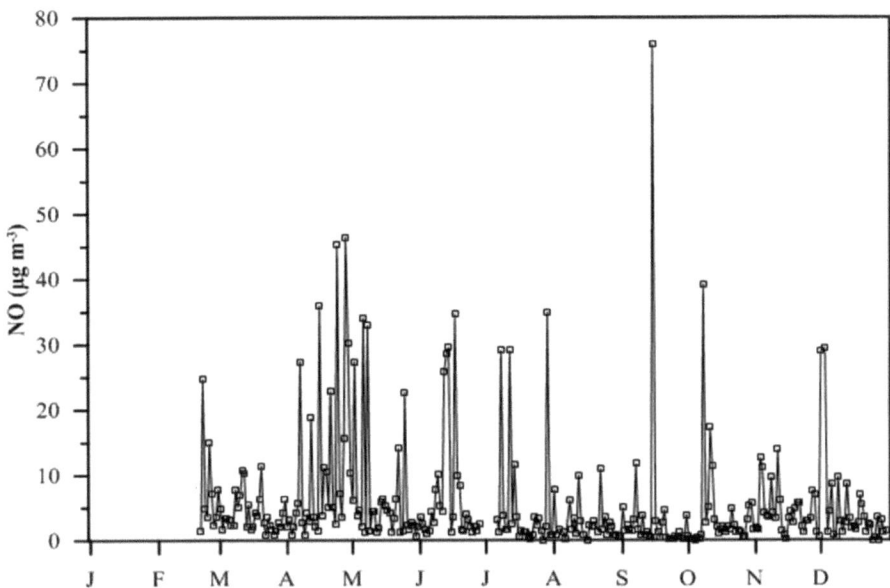

Fig. 14 Daily maximum NO levels during 2013 (derived from maximum 8-h values)

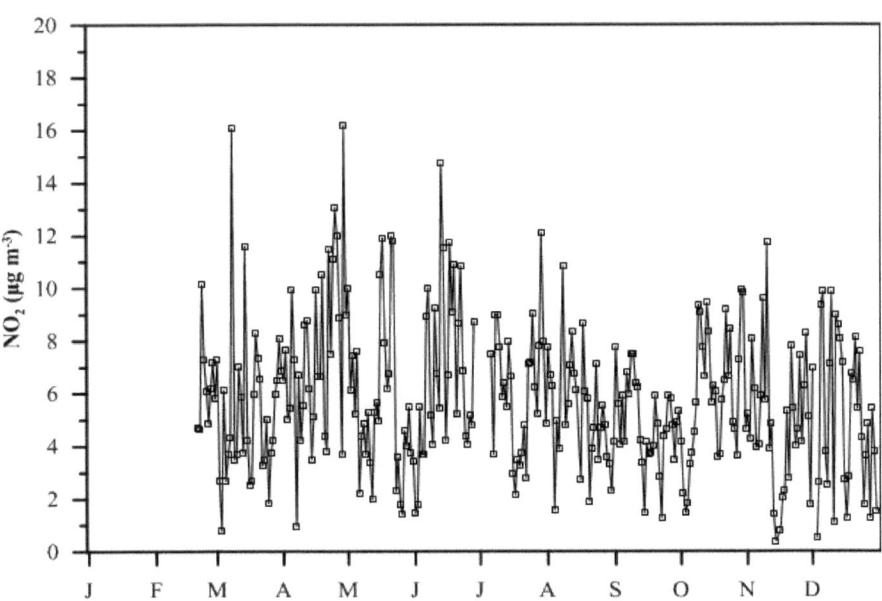

Fig. 15 Daily mean NO₂ concentrations during 2013 (derived from mean 8-h values)

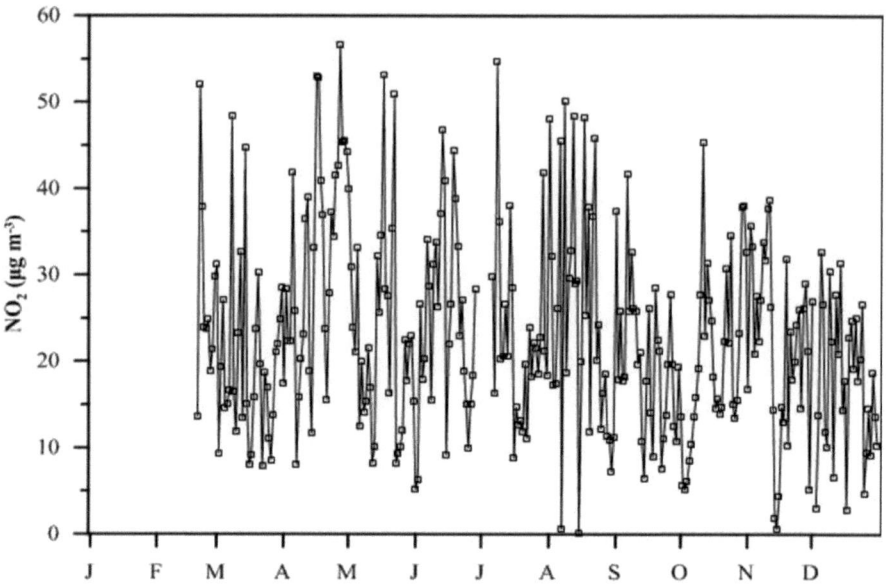

Fig. 16 Daily maximum NO$_2$ levels during 2013 (derived from max. 8-h values)

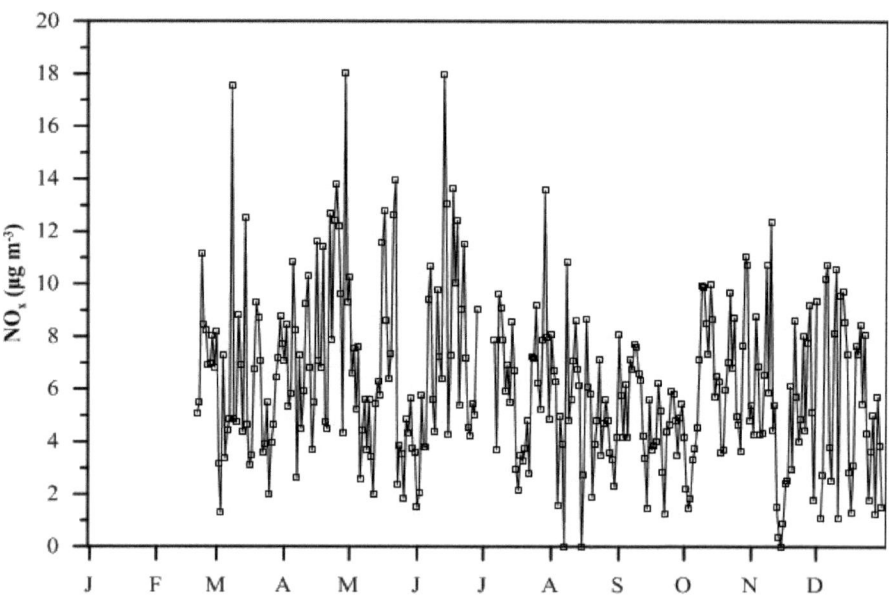

Fig. 17 Daily mean NO$_x$ concentrations during 2013 (derived from mean 8-h values)

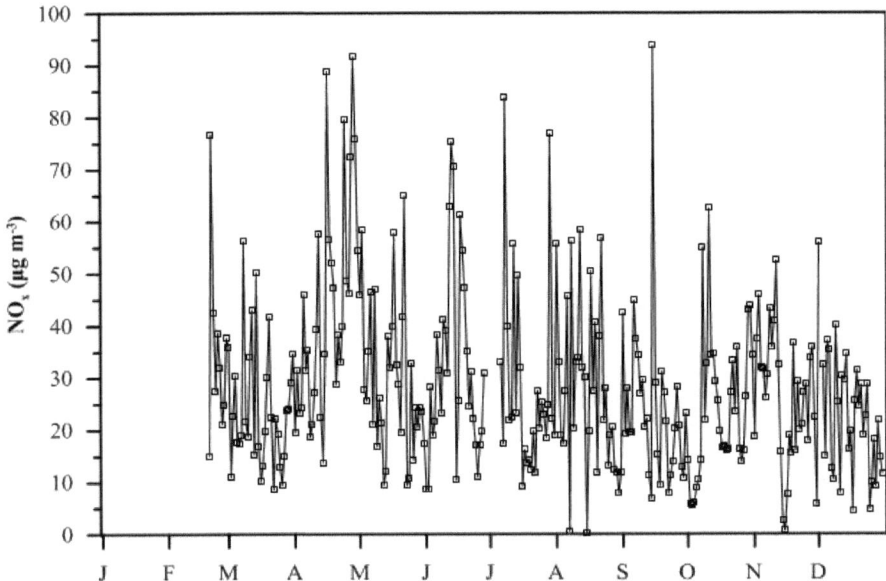

Fig. 18 Daily maximum NO$_x$ levels during 2013 (derived from max. 8-h values)

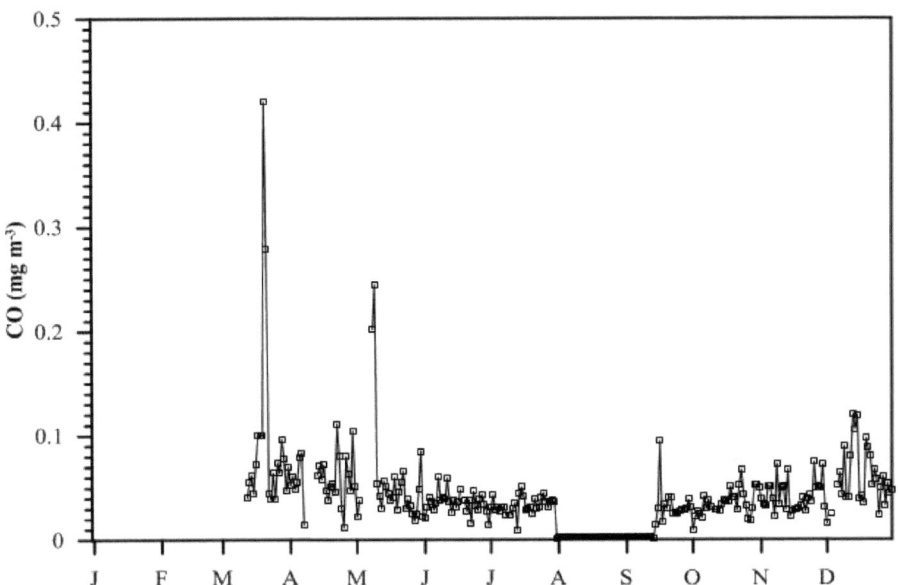

Fig. 19 Daily mean CO concentrations during 2013, measured at the UPC site

The variation of daily mean CO concentrations during the measurement period in the year 2013, measured at the UPC site, is presented in Fig. 19. These daily mean values were calculated as the mean value of the related 8-h average values recorded. These CO levels are much less than the limit value of 10 mg m^{-3} (Directive 2008/50/EC). No any limit exceedance was detected.

3.2 Variation of monthly mean pollutant concentrations

The variation of monthly mean concentration values of PM_{10}, $PM_{2.5}$ and PM_1 during the measurement period in years 2012 and 2013, measured at the UPC site, are presented in Fig. 20 and 21, respectively. The contribution of natural sources, as transportation of Sahara's dust, forest fires, seawater aerosols etc., confuses the justification of this variation.

Higher O_3 values are recorded at the UPC site during the warm period of the year 2013 (March to August), compared to the cold period of the year (Fig. 22). This variation depends on the parameters affecting ozone's formation, such as intensity and duration of sunshine, which are increased during the warm period. The present findings are comparable to those observed by Gerasopoulos *et al.* (2006) at Finokalia's monitoring station of air pollution.

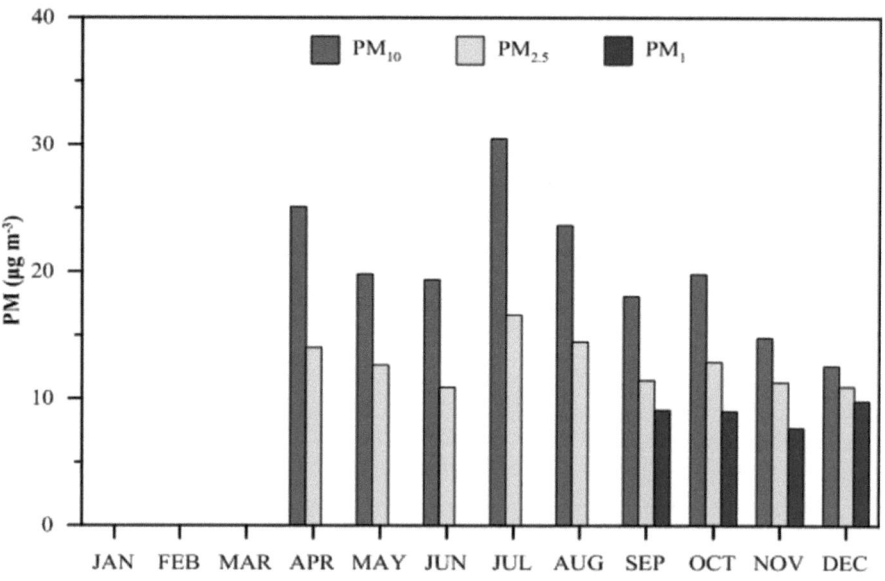

Fig. 20 Monthly mean PM_{10}, $PM_{2.5}$ and PM_1 concentrations during 2012

26

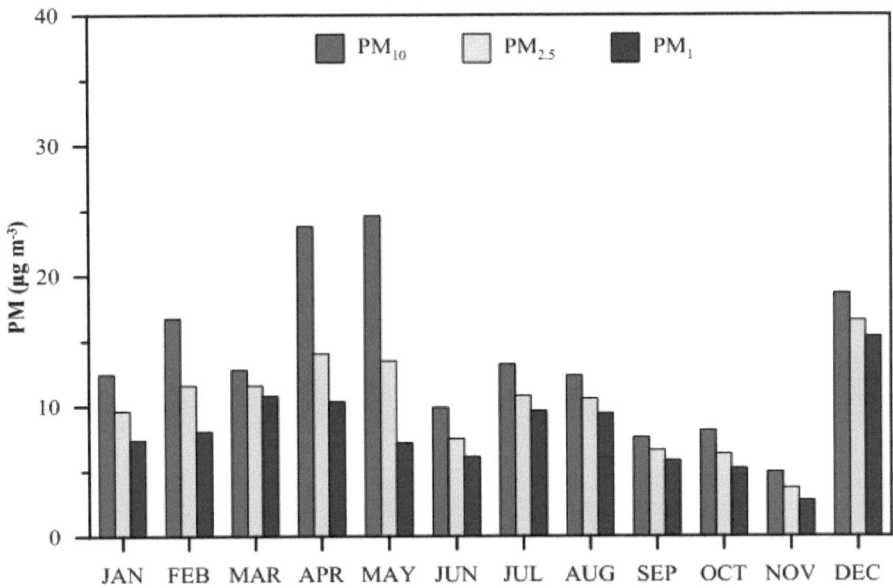

Fig. 21 Monthly mean PM_{10}, $PM_{2.5}$ and PM_1 concentrations during 2013

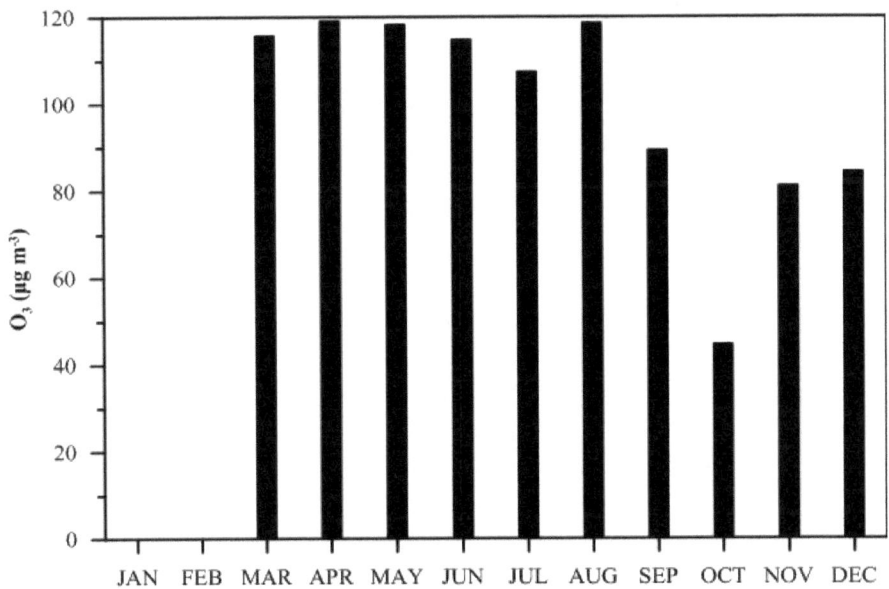

Fig. 22 Monthly mean O_3 concentrations during 2013, measured at the UPC site

The variation of mean and averaged maximum monthly concentration values of NO, NO_2 and NO_x during the measurement period in the year 2013, measured at the UPC site, are presented in Fig. 23 and 24, respectively. These values are much lower than the annual NO_x limit of 30 $\mu g\ m^{-3}$, which is set for the protection of vegetation (Directive 2008/50 / EC).

The variation of monthly mean CO concentration values during the measurement period in the year 2013, measured at the UPC site, is presented in Fig. 25. Reduced CO concentrations are observed during the summer period, especially in August, because of insignificant or zero traffic load at UPC.

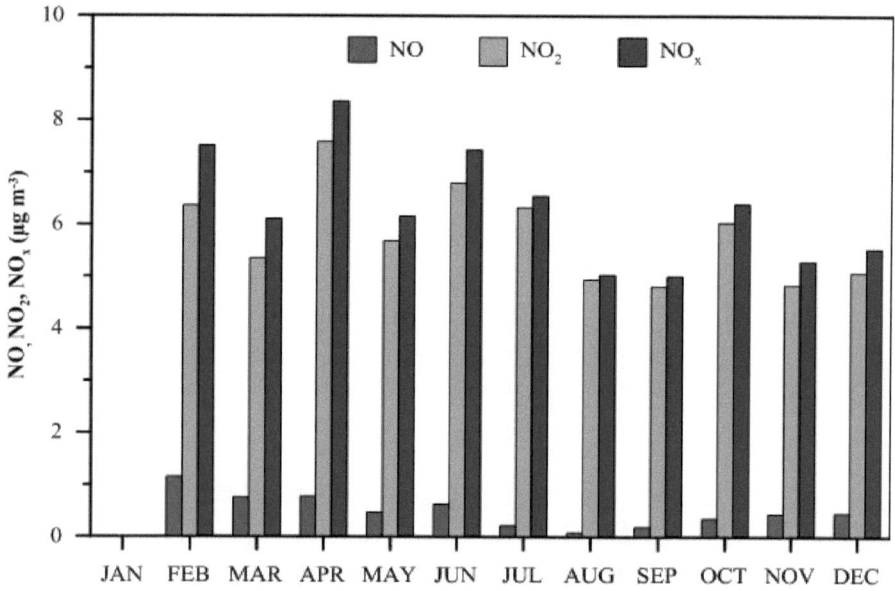

Fig. 23 Monthly mean NO, NO_2 and NO_x concentrations during 2013 (derived from mean 8-h values)

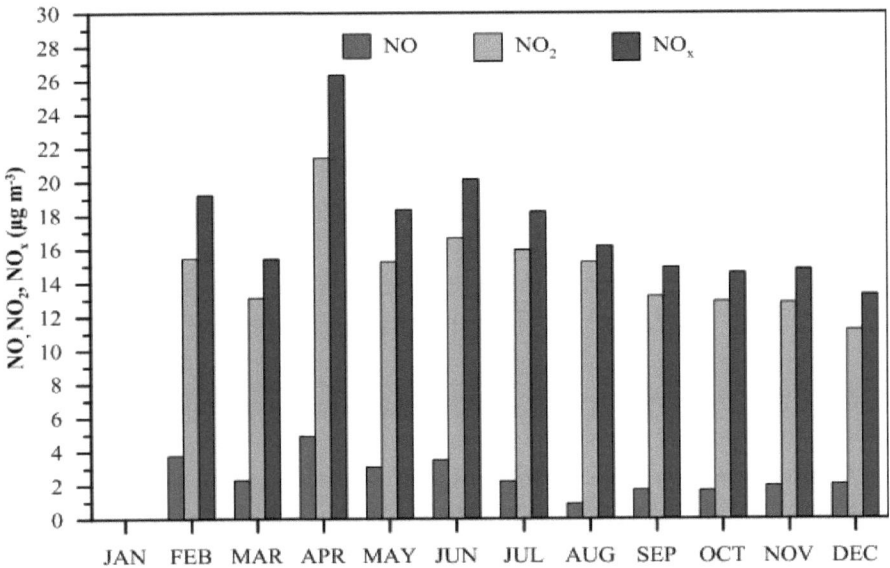

Fig. 24 Monthly mean maximum NO, NO_2 and NO_x concentrations during 2013 (derived from maximum 8-h values)

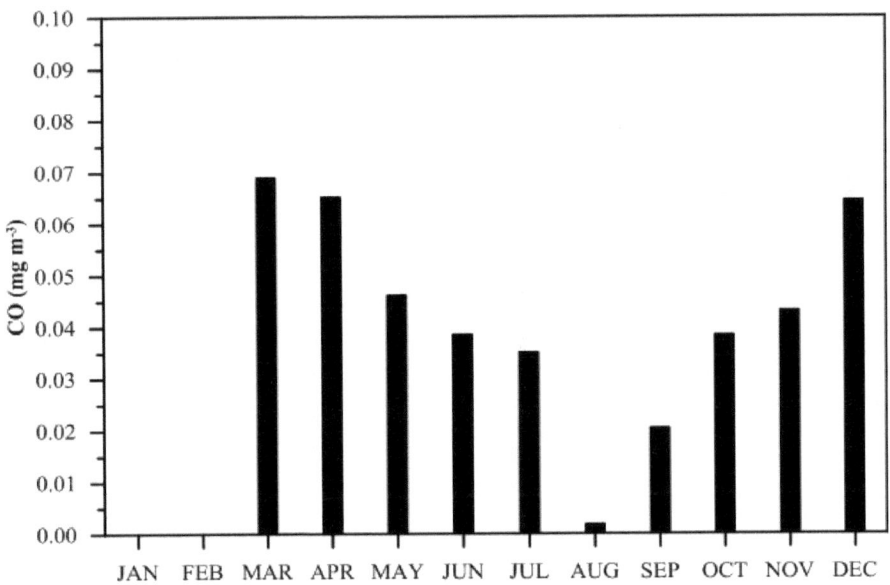

Fig. 25 Monthly mean CO concentrations during 2013, measured at the UPC site

3.3 Weakly variation of mean pollutant concentrations

The weakly variation of daily averaged concentration values for PM_{10}, $PM_{2.5}$ and PM_1 during the measurement period in the years 2012 and 2013, measured at the UPC site, are presented in Fig. 26 and 27, respectively. No specific change has been noticed between the two years, regarding the day of the week.

The weakly variation of daily averaged O_3 concentration during the measurement period in the year 2013, measured at the UPC site, is presented in Fig. 28. The stability of O_3 concentrations throughout the typical week is justified due to the mechanism of O_3 formation, which is rather weather dependent than anthropogenic.

The weakly variation of averaged and mean maximum daily concentration values of NO, NO_2 and NO_x during the measurement period in the year 2013, measured at the UPC site, is shown in Fig. 29 and 30, respectively. It is observed that the typical daily values depend on both anthropogenic activities and natural conditions.

The weakly variation of averaged daily CO concentration values during the measurement period in the year 2013, measured at the UPC site, is shown in Fig. 31, which indicates very low levels with a slight decrease during the typical weekend, when traffic activities are also reduced.

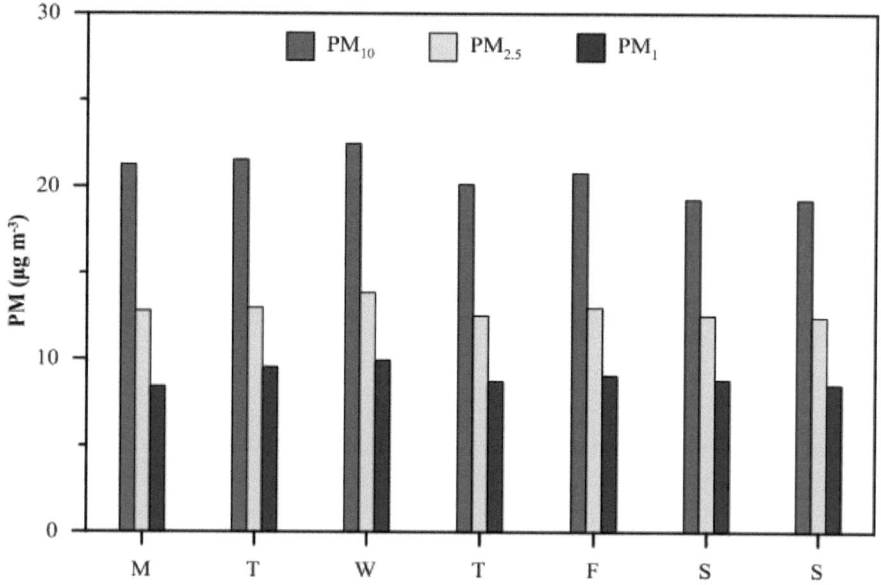

Fig. 26 Weakly mean PM_{10}, $PM_{2.5}$ and PM_1 concentrations during 2012

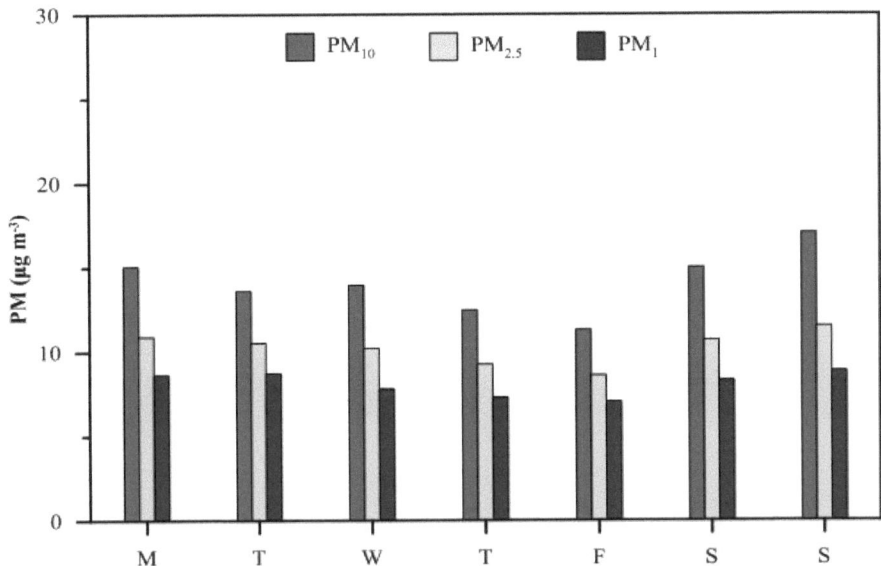

Fig. 27 Weakly mean PM_{10}, $PM_{2.5}$ and PM_1 concentrations during 2013

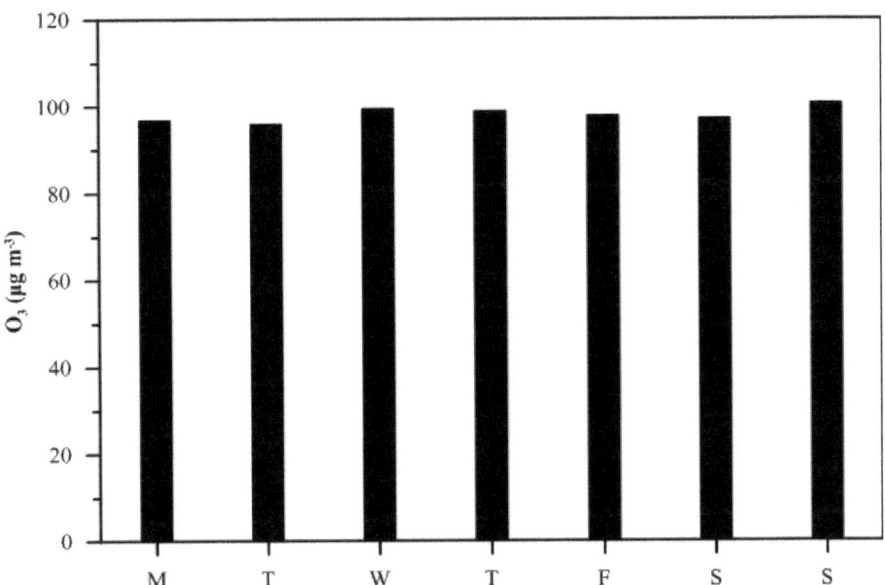

Fig. 28 Weakly mean O_3 concentrations during 2013, measured at the UPC site

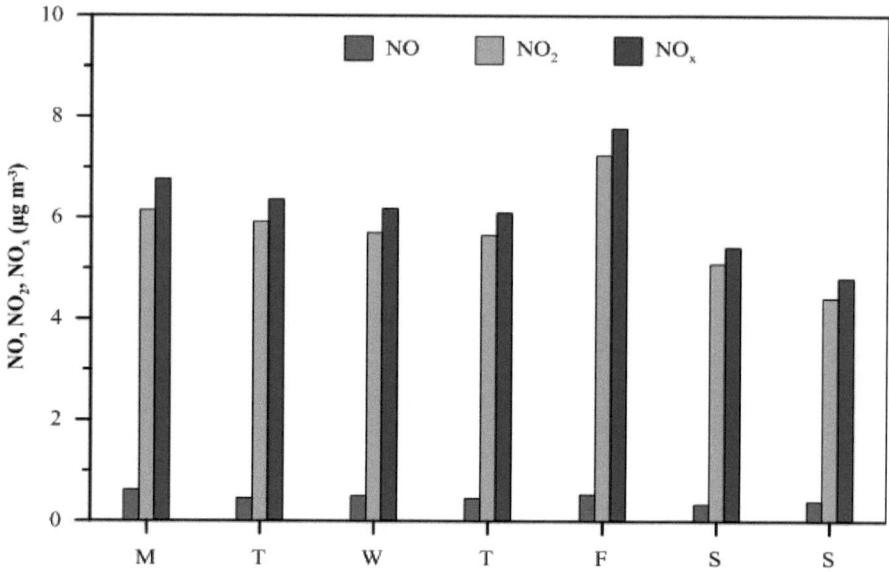

Fig. 29 Weakly mean NO, NO$_2$ and NO$_x$ concentrations during 2013 (derived from mean 8-h values)

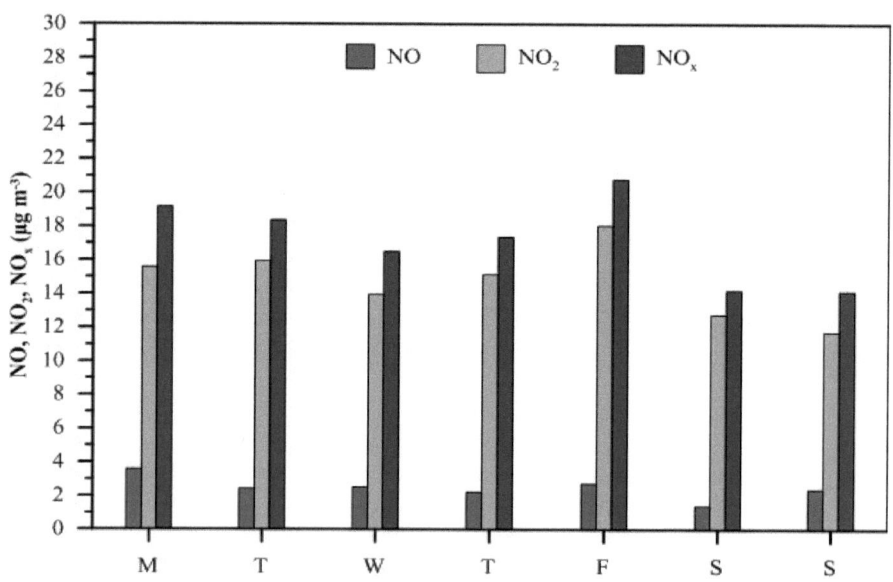

Fig. 30 Weakly mean maximum NO, NO$_2$ and NO$_x$ concentrations during 2013 (derived from maximum 8-h values)

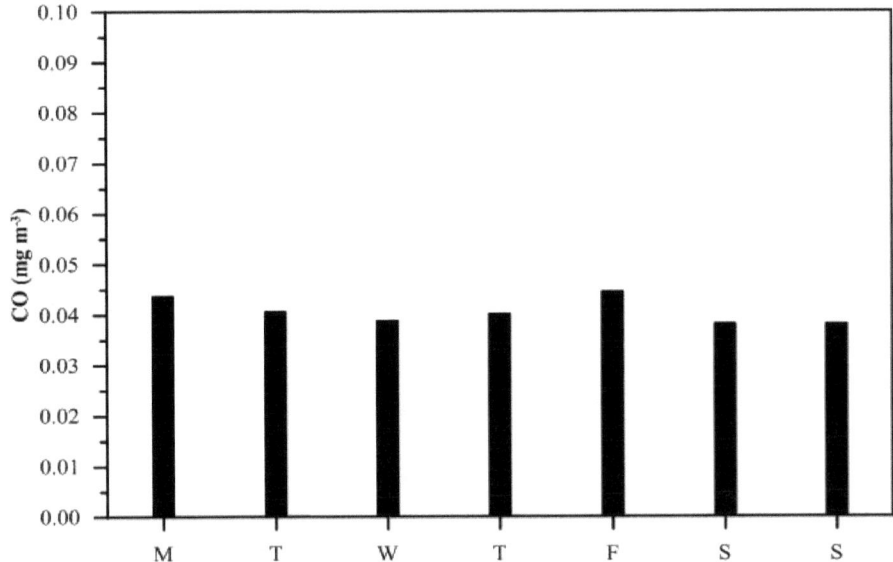

Fig. 31 Weakly mean CO concentrations during 2013, measured at the UPC site

3.4 Diurnal variation of mean hourly pollutant concentrations

The diurnal variation of hourly mean concentrations of selected pollutants in the area of UPC is presented. Figures 32 and 33 show the diurnal variation of PM_{10}, $PM_{2.5}$ and PM_1 during 2012 and 2103. In general, the PM values presented one peak during morning and another one during evening hours. This pattern agrees with the typical pattern of primary air pollutants, justifying that PM_{10}, $PM_{2.5}$ and PM_1 at the UPC site behave as rather primary than secondary air pollutants.

The diurnal variation of mean hourly O_3 values during 2013, measured at the UPC site, is presented in Fig. 34. The concentrations are low during early in the morning and are increasing in the afternoon with a peak of 110 μg m^{-3}, approximately, in the time interval 18:00 – 21:00, which agrees with the general pattern for secondary air pollutants. Similar behavior is observed by Gerasopoulos *et al.* (2006) for the diurnal ozone cycle at Finokalia of Crete, Greece, during the period 1997-2004 with a peak value 100 μg m^{-3}, approximately.

The diurnal variation of NO_x, measured at the UPC site, is presented in Fig. 35 and 36 during 2013 for the 8-h mean values and 8-h maximum values, respectively. NO, NO_2 and NO_x concentration levels are very low displaying their maximum levels at 08:00 to 16:00.

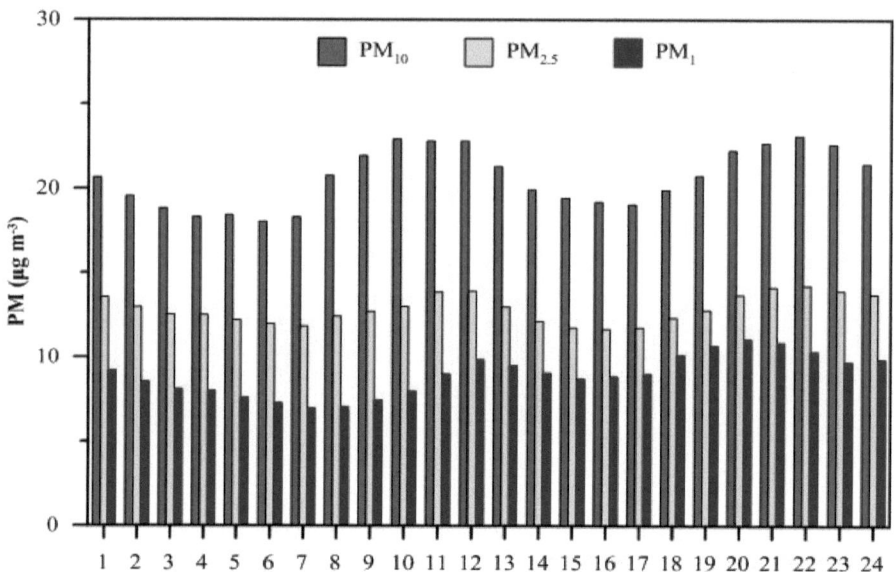

Fig. 32 Diurnal variation of mean hourly PM_{10}, $PM_{2.5}$ and PM_1 levels during 2012

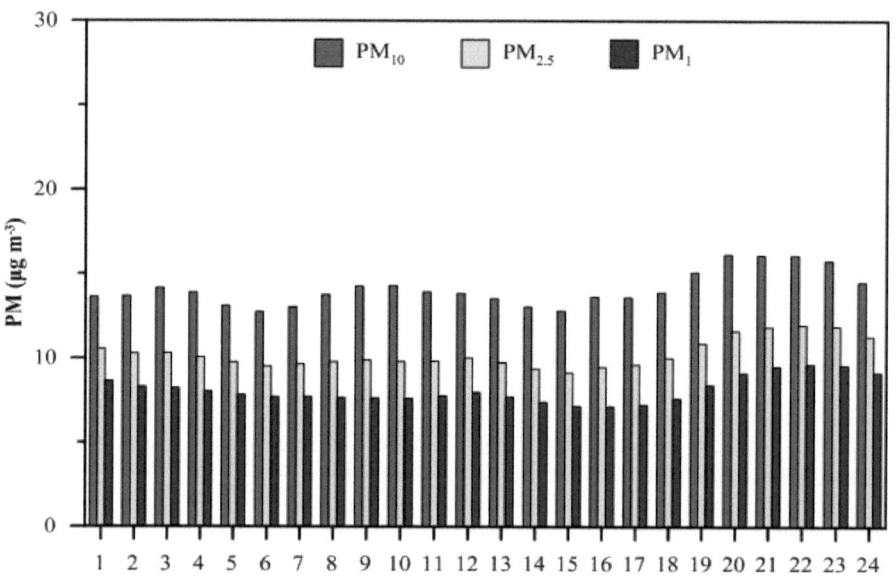

Fig. 33 Diurnal variation of mean hourly PM_{10}, $PM_{2.5}$ and PM_1 levels during 2013

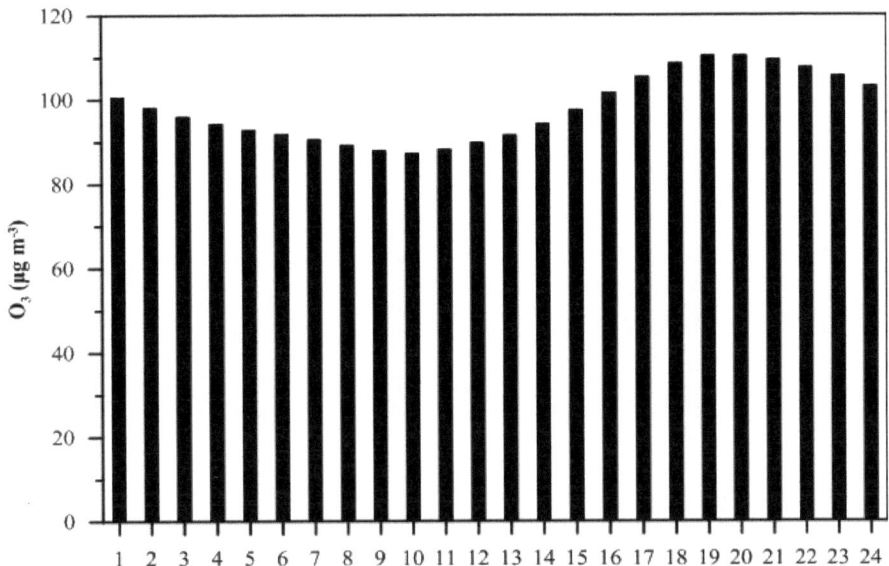

Fig. 34 Diurnal variation of mean hourly O_3 concentrations during 2013

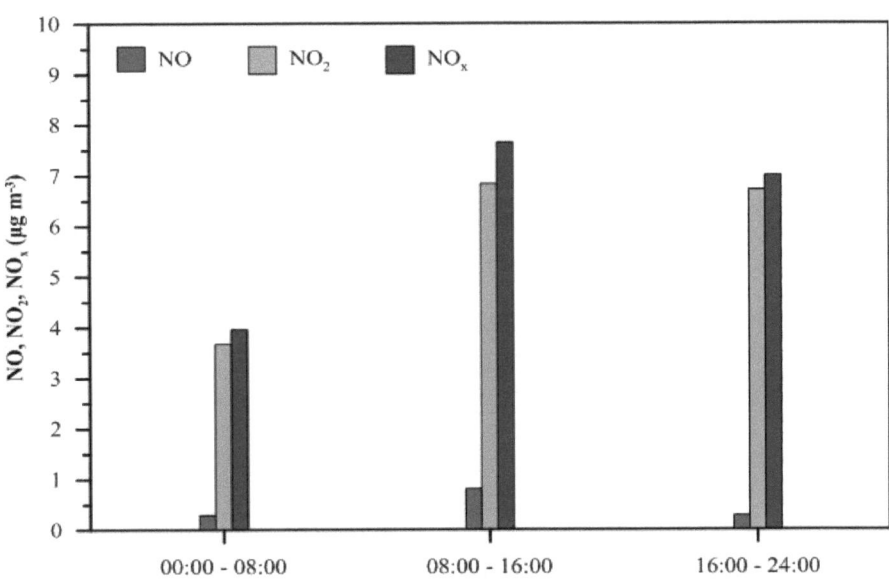

Fig. 35 Diurnal variation of mean 8-h NO, NO_2 and NO_x concentrations during 2013
(derived from mean 8-h values)

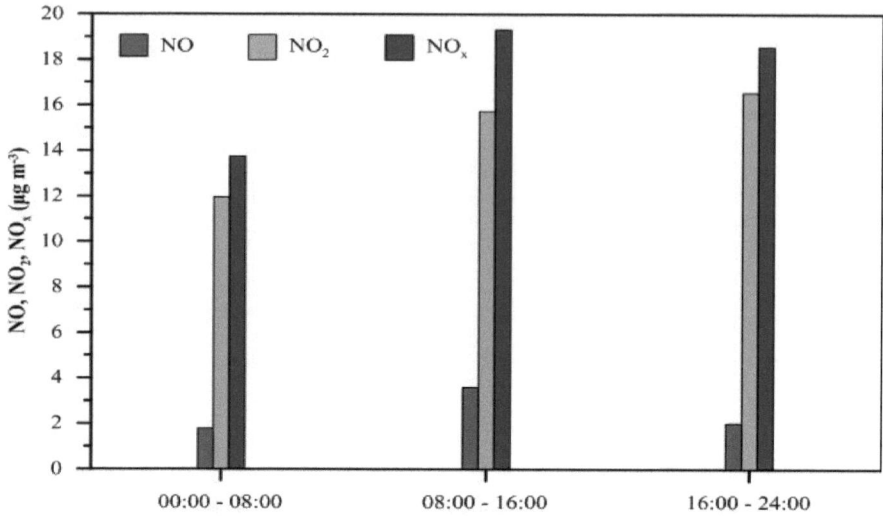

Fig. 36 Diurnal variation of maximum 8-h NO, NO_2 and NO_x concentrations during 2013 (derived from maximum 8-h values)

The diurnal variation of CO concentrations during 2013, measured at the UPC site, is presented in Fig. 37. The values decrease during night hours, becoming low early in the morning and then increase during the day, following the traffic pattern, which is somewhat different at UPC compared to the traffic pattern of urban areas, due to differences between the work programs.

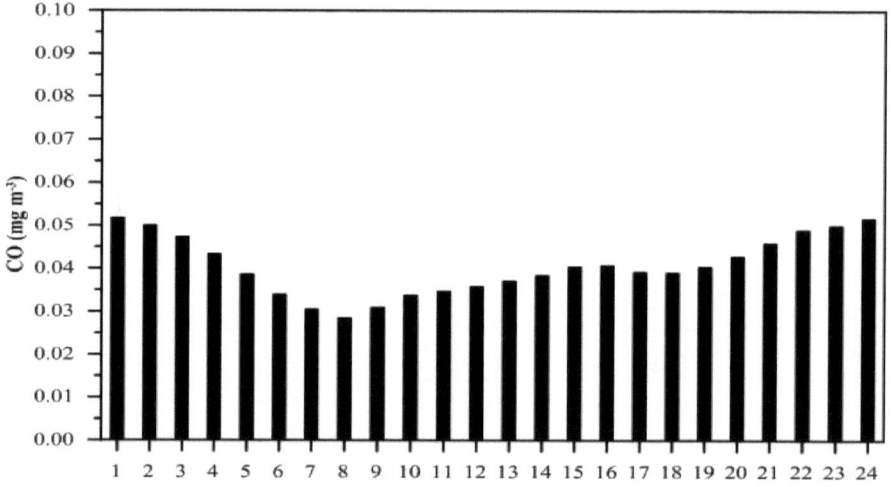

Fig. 37 Diurnal variation of mean CO concentrations during 2013

36

4. STATISTICAL DATA OF POLLUTANTS AND COMPARISONS WITH DATA MEASURED IN OTHER HELLENIC SITES

The statistical data for PM_{10}, $PM_{2.5}$, PM_1, O_3, NO, NO_2 and NO_x concentrations during the measurement period of years 2012 and 2013 at the University of Patras Campus (UPC) are shown in Table 3. At the upper side of Table 3 are the statistical data of the daily mean values. At the second section of Table 3, the statistical data of hourly mean concentrations for PM_{10}, $PM_{2.5}$, PM_1 and O_3 are presented and also the statistical data for 8-h mean values of NO, NO_2 and NO_x.

A summary of the statistical data of the mean annual values for PM_{10}, $PM_{2.5}$, PM_1, O_3, NO, NO_2 and NO_x at UPC for the measurement period in comparison to the available corresponding values of other monitoring stations in Greece, which are installed and operated from the "Greek National Monitoring Network of Atmospheric Pollution (GNMNAP)" (http://www.ypeka.gr), is given in Tables 4 and 5. In case of unavailable concentration values for the years of air pollution monitoring at UPC, the year that data come from is noted in brackets.

Table 3. Statistical data of air pollutant concentrations ($\mu g\ m^{-3}$) at the UPC site

	PM_{10}		$PM_{2.5}$		PM_1		O_3	CO	NO	NO_2	NO_x
Sampling Period	07/04 – 31/12/2012	2013	07/04 – 31/12/2012	2013	07/09 – 31/12/2012	2013	12/03 – 31/12/2013	11/03- 31/12/2013	19/02 – 31/12/2013		
	Daily Values										
Minimum	0.7	1.4	0.6	1.0	0.4	0.7	25.2	0.8	0.0	0.0	0.0
Maximum	91.3	98.7	50.8	42.5	21.7	23.9	176.9	420.3	3.2	16.2	18.0
Average	20.4	14.0	12.7	10.2	8.8	8.1	116.2	41.0	0.5	5.7	6.2
Median	18.9	11.5	12.5	9.2	8.6	7.5	123.0	36.7	0.3	5.4	5.8
Standard Deviation	10.8	12.0	5.8	6.1	5.5	4.4	29.9	39.7	0.5	2.8	3.1
98% values <	45.9	53.5	27.5	26.1	21.2	20.1	165.0	119.7	1.9	12.0	13.6
Completion (%)	69.7	76.4	69.7	76.4	27.9	76.4	76.4	77.3	84.1	84.1	84.1
	Hourly Values								8-h Values		
Minimum	0.2	0.3	0.2	0.38	0.0	0.2	7.8	1.9	0.0	0.0	0.0
Maximum	386.5	368.1	157.0	143.9	58.0	53.0	186.3	327.8	8.1	31.0	33.7
Average	20.6	14.1	12.9	10.3	9.0	8.1	98.1	40.6	0.5	5.7	6.2
Median	18.6	11.1	12.1	8.8	7.8	7.2	99.9	34.8	0.1	4.8	5.2
Standard Deviation	14.5	16.2	7.5	8.0	6.8	5.4	33.0	35.2	0.8	4.3	4.8
98% values <	53.1	52.6	31.1	29.8	27.1	22.7	154.8	136.3	3.1	17.0	19.0
Completion (%)	67.5	73.8	67.5	73.8	26.2	73.8	72.8	74.9	84.1	84.1	84.1

37

Table 4. Mean annual data for air pollutants at the UPC and other Hellenic sites during 2012

City Area	Patras		Alliartos	Athens		Piraeus	Ag. Paraskevi	Thessaloniki		Larisa	Volos	Crete	Ioannina
Station Location	UPC	Patras (Patra-1)	Aliartos (ALI)	Lykovrissi (LYK)	Patission (PAT)	Piraeus (PIR)	Ag. Paraskevi (AGP)	Ag. Sofia (AGS)	Kalamaria (KAL)	Larisa (LAR)	Volos (VOL)	Irakleio (IRA)	Ioannina (IOA)
Longitude	21° 47' 22''	21° 44' 18''	23° 06' 37''	23° 47' 20''	23° 43' 59''	23° 38' 43''	23° 49' 10''	22° 56' 43''	22° 57' 33''	22° 27' 12''	22° 56' 35''	25° 04' 48''	20° 51' 00''
Latitude	38° 17' 22''	38° 15' 11''	38° 22' 31''	38° 04' 04''	37° 59' 58''	37° 56' 41''	37° 59' 42''	40° 38' 02''	40° 34' 44''	39° 40' 03''	39° 21' 59''	35° 19' 57''	39° 37' 12''
Altitude (m)	61	16	110	234	105	4	290	27	60	15	31	10	485
Area type	Background	Traffic	Background	Suburban	Traffic	Traffic	Background	Traffic	Background	Traffic	Background	Traffic	Background
Station type	Suburban	Urban	Rural	Suburban	Urban	Urban	Suburban	Urban	Suburban	Urban	Urban	Urban	Urban
O_3 ($\mu g\ m^{-3}$)		80 [2011] (58)[1]	52	72 (43)[1,2]	24 (0)[1]	41 (0)[1]	85 (85)[1,2]	55 (37)[1]		33 [2011] (0)[1]			18 [2011] (0)[1]
NO ($\mu g\ m^{-3}$)				14	77	35	3						
NO_2 ($\mu g\ m^{-3}$)		20 [2011] (0)[4]	2.74 (0)[4]	21 (0)[4]	64[3] (0)[4]	41[3] (0)[4]	9 (0)[4]	24 (0)[4]	32 [2009] (0)[4]	21 [2011] (0)[4]			15 [2010] (0)[4]
SO_2 ($\mu g\ m^{-3}$)		38 [2011] (0)[5]	2.0 (0)[5]		7 (0)[5]	10 (0)[5]							
PM_{10} ($\mu g\ m^{-3}$)	20.4 (4)[7]	37 [2010] (34)[7]	29 [2011] (0)[7]	37 (40)[7]		39 (38)[7]	23 (3)[7]	41[6] (79)[7]		33 [2011] (29)[7]	31 (8)[7]		28 [2010] (18)[7]
$PM_{2.5}$ ($\mu g\ m^{-3}$)	12.7			22		27 (2011)[8]	15						
CO ($mg\ m^{-3}$)		0.6 [2011] (0)[9]			1.5 (0)[9]	0.8 (0)[9]	0.8 (0)[9]	0.8 (0)[9]	0.5 (0)[9]	0.4 (0)[9]			

[1] Exceedances of the max. daily 8-h limit of 120 µg m⁻³
[2] Exceedances during 2010-2012
[3] Exceeds the annual limit value of 40 µg m⁻³
[4] Exceedances of the hourly limit value of 200 µg m⁻³
[5] Exceedances of daily limit of 125 µg m⁻³
[6] Exceeds the annual limit value of 40 µg m⁻³
[7] Exceedances of the daily limit of 50 µg m⁻³
[8] Exceeds the annual target value of 25 µg m⁻³ (2015)
[9] Number of exceedances of the max. daily 8-h limit of 10 mg m⁻³

Source: http://www.ypeka.gr

Table 5. Mean annual data of air pollutant concentrations at the UPC and other Hellenic sites during 2013

City Area	Patras	Alliartos	Athens			Thessaloniki			Larisa	Volos	Crete	Ioannina
Station	UPC	Aliartos (ALI)	Lykovrissi (LYK)	Patission (PAT)	Piraeus (PIR)	Ag. Paraskevi (AGP)	Ag. Sofia (AGS)	Kalamaria (KAL)	Larisa (LAR)	Volos (VOL)	Irakleio (IRA)	Ioannina (IOA)
Geog. Longitude	21° 47' 22''	23° 06' 37''	23° 47' 20''	23° 43' 59''	23° 38' 43''	23° 49' 10''	22° 56' 43''	22° 57' 33''	22° 27' 12''	22° 56' 35''	25° 04' 48''	20° 51' 00''
Geog. Latitude	38° 17' 22''	38° 22' 31''	38° 04' 04''	37° 59' 58''	37° 56' 41''	37° 59' 42''	40° 38' 02''	40° 34' 44''	39° 40' 03''	39° 21' 59''	35° 19' 57''	39° 37' 12''
Altitude (m)	61	110	234	105	4	290	27	60	15	31	10	485
Area Type	Background	Background	Suburban	Traffic	Traffic	Background	Traffic	Background	Traffic	Background	Traffic	Background
Station Type	Suburban	Rural	Suburban	Urban	Urban	Suburban	Urban	Suburban	Urban	Urban	Urban	Urban
O$_3$ (µg m^{-3})	98.4 (148)[1]	66 (59)[1,2]	25 (0)[1]	33 (0)[1]	83 (76)[1,2]	48 (34)[1]		69 (58)[3]				
NO (µg m^{-3})	0.5	15	83	40	3							
NO$_2$ (µg m^{-3})	5.7 (0)[5]	21 (0)[5]	52[4] (0)[5]	36 (0)[5]	8 (0)[5]	21 (0)[5]		34 (0)[5]	31[9] (0)[5]			
SO$_2$ (µg m^{-3})	1.5 (0)[6]		7 (0)[6]	10 (0)[6]				13 (0)[6]				
PM$_{10}$ (µg m^{-3})	14.0 (7)[7]	42 (70)[7,8]		37 (39)[7,8]	26 (19)[7]	22 (28)[7]	31	39 (31)[7]	38 (59)[7,8]	33 (37)[7,8]		
PM$_{2.5}$ (µg m^{-3})	10.2	12[9]			10							
CO (mg m^{-3})	0.041		1.4 (0)[10]	0.8 (0)[10]		0.4 (0)[10]	0.4 (0)[10]	0.6 (0)[10]	0.3 (0)[10]			

[1] Exceedances of the max. daily 8-h limit of 120 µg m^{-3}
[2] Exceedances during 2010-2012
[3] Except 2012
[4] Exceeds the annual limit value of 40 µg m^{-3}
[5] Exceedances of the hourly limit of 200 µg m^{-3}
[6] Exceedances of daily limit of 125 µg m^{-3}
[7] Number of exceedances of the daily limit of 50 µg m^{-3}
[8] More than 35 exceedances during the year
[9] Few sample data
[10] Exceedances of the max. daily 8-h limit of 10 mg m^{-3}
Source: http://www.ypeka.gr

The monthly concentration values of atmpospheric pollutants at the UPC site are compared to the corresponding values available from other air quality monitoring sites in Greece. The monitoring sites and the corressponding data are provided by the "Greek National Monitoring Network of Atmospheric Pollution (GNMNAP)" (http://www.ypeka.gr).

The monthly mean PM_{10} concentrations during 2012-2013 monitored at UPC are shown in Fig. 38 in comparison with Piraeus (PIR), Agia Paraskevi (AGP), Patission (PAT) and Lykovrissi (LYK) corresponding data. Generally, it is obvious that PM_{10} variation at UPC is almost the lowest, showing tha PM_{10} levels at UPC could be characterized as background levels.

The monthly mean $PM_{2.5}$ concentrations during 2012 - 2013 at UPC are shown in Fig. 39 in comparison with Piraeus (PIR), Agia Paraskevi (AGP) and Lykovrissi (LYK). The variation of the present data follows the variation of AGP data. The site of AGP is characterized as background – suburban.

The monthly mean O_3 concentrations during 2012 - 2013 in UPC is presented in Fig. 40 in comparison with Piraeus (PIR), Agia Paraskeyi (AGP), Patission (PAT), Lykovrissi (LYK) and Aliartos (ALI). Again, O_3 variation at UPC follows the variations of the background – suburban sites (AGP and LYK), which are affected by the highest levels of secondary air pollution.

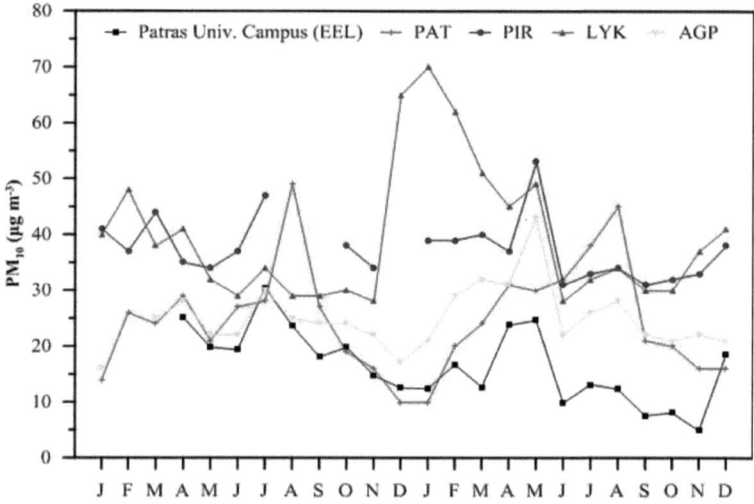

Fig. 38 Monthly mean PM_{10} concentrations during 2012-2013 in UPC and other monitoring stations in Greece

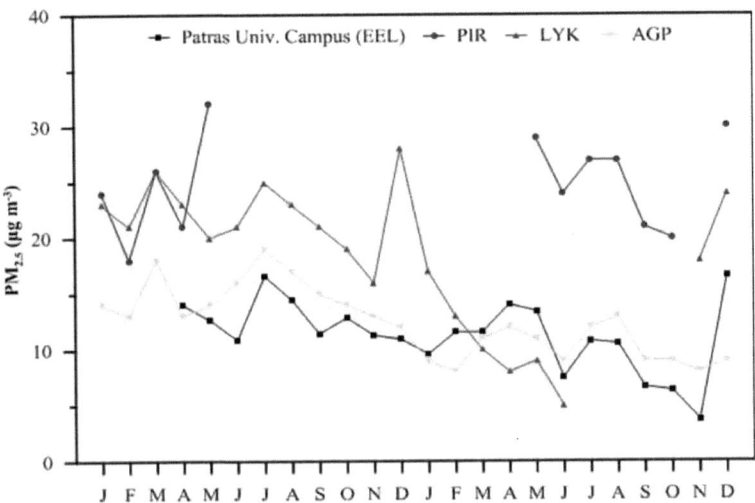

Fig. 39 Monthly mean $PM_{2.5}$ concentrations during 2012-2013 in UPC and other monitoring stations in Greece

In the next two figures (Fig. 41 and 42), the monthly variation of NO and NO_2 during 2012 – 2013 is presented. The variation of UPC is similar to that of AGP

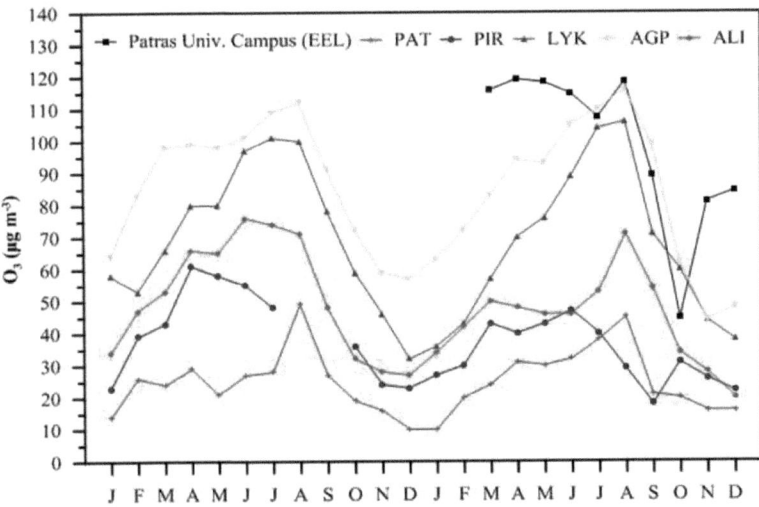

Fig. 40 Monthly mean O_3 concentrations during 2012-2013 in UPC and other monitoring stations in Greece

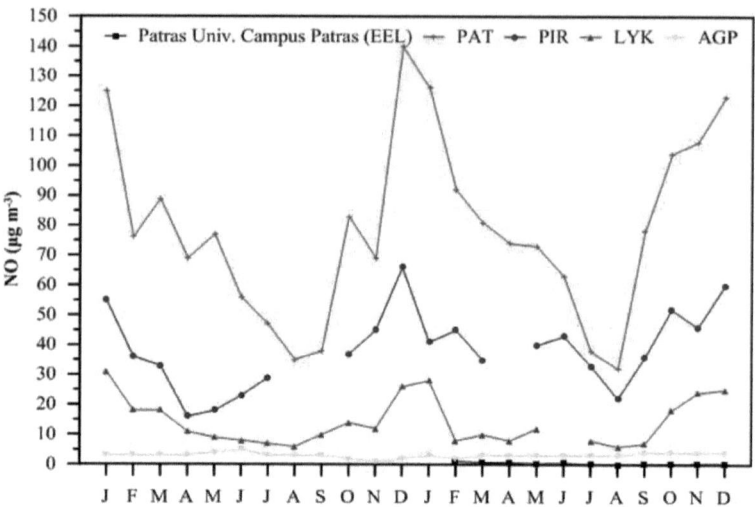

Fig. 41 Monthly mean NO concentrations during 2012-2013 in UPC and other monitoring stations in Greece

for both NO (Fig. 41) and NO_2 (Fig. 42). Also, Fig. 42 indicates that the NO_2 variation at the UPC site is comparable to that in AGP and ALI sites.

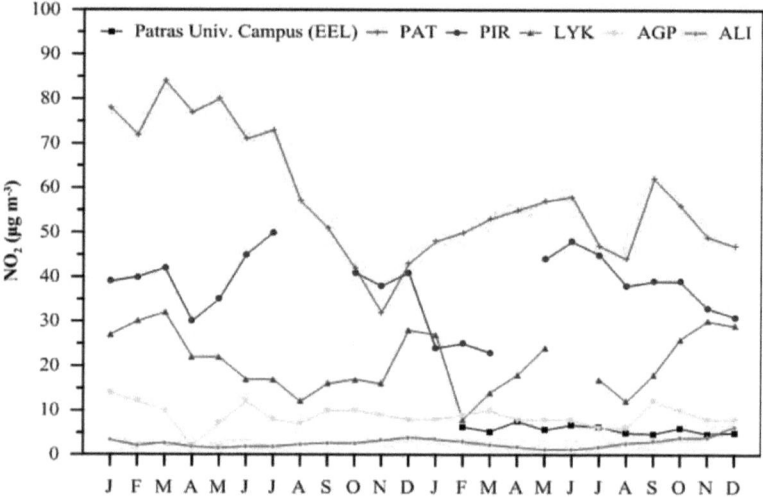

Fig. 42 Monthly mean NO_2 concentrations during 2012-2013 in UPC and other monitoring stations in Greece

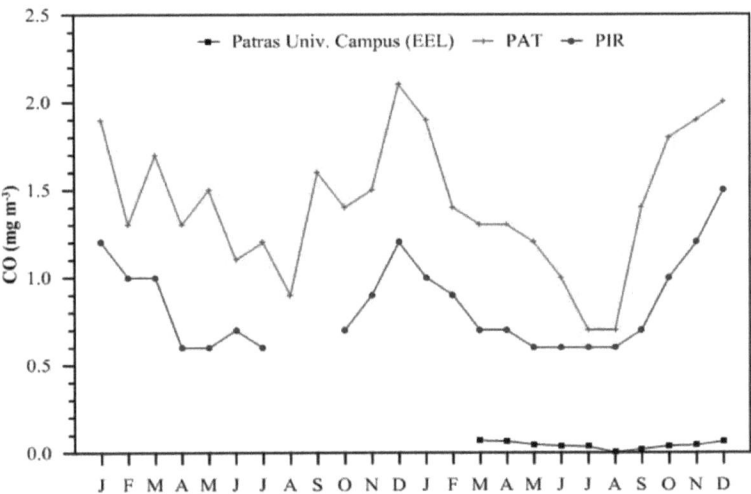

Fig. 43 Monthly mean CO concentrations during 2012-2013 at UPC and other monitoring stations in Greece

The monthly CO variation at the UPC site is shown in Fig. 43 in comparison to the variation in PAT and PIR sites. The low CO levels monitored at UPC compared to that of PAT and PIR sites are due to corresponding traffic loads.

The diurnal variation of hourly mean O_3 concentrations at UPC compared to available data for other Greek cities is shown in Fig. 44, where O_3 levels at UPC follow well enough the pattern of the background site of Finocalia, Crete. The UPC diurnal variation of hourly mean O_3 levels agree also with similar observations during November 1992 and January 1993 made by Danalatos and Glavas (1996).

Paying attention on Fig. 44, although one may see that the UPC O_3 diurnal variation follows approximately the average of all the curves, there are circumstances, when this diurnal variation exceeds the corresponding curves for other areas. This observation reveals that, although the process of formation of the secondary air pollutants in Patras is rather moderate, however there exist some extraordinary events of O_3 formation in the afternoon hours, as in April 1992 (Danalatos and Glavas 1996). This finding highlights the necessity of secondary air pollution monitoring, as it has also pointed out by Yannopoulos (2014).

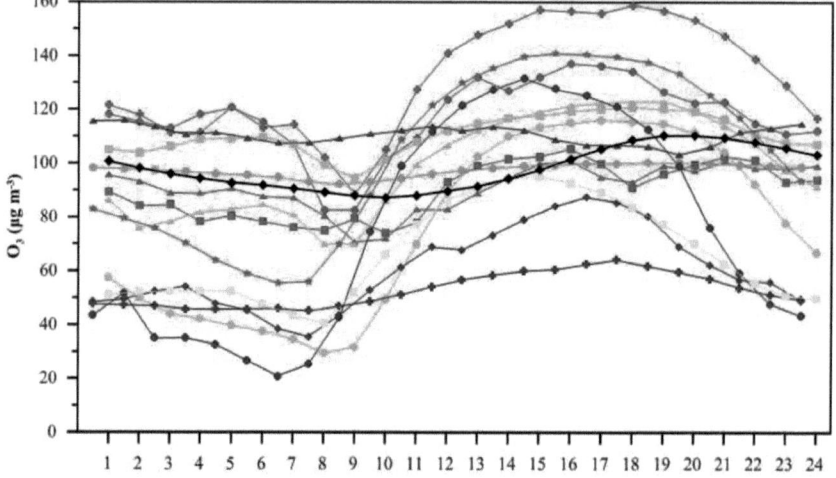

Fig. 44 Diurnal variation of hourly mean O_3 concentrations at the UPC site compared to available data from Greece

44

5. COMPARISON WITH LIMIT VALUES AND VALUES MEASURED IN SEVERAL MEDITERRANEAN SITES OF EUROPE

The statistical data for the concentrations of air pollutants measured by the EEL Station at the site of the University of Patras Campus (UPC) are compared in Table 6 to the corresponding data of other Hellenic sites of several types (background of traffic sites and rural, suburban or urban sites). It is observed that both the primary and secondary air quality characteristics measured at UPC compare well to the background suburban site of Ag. Paraskevi, Athens.

Table 7 includes comparisons between statistical data for the daily concentration values of air pollutants measured by the EEL Station at UPC and other Mediterranean sites of Europe of approximately same latitude (http://www.eea.europa.eu/data-and-maps/data/airbase-the-european-air-quality-database-8). Measurements show that both the primary and secondary air quality characteristics at UPC site are comparable to the corresponding data monitored at rural or suburban sites of European Mediterranean countries.

Table 8 includes comparisons between statistical data for the annual concentration values of air pollutants measured by the EEL Monitoring Station at the UPC and other Mediterranean sites of Europe of approximately same latitude (http://www.eea.europa.eu/data-and-maps/data/airbase-the-european-air-quality-database-8). Measurements show again that both the primary and secondary air quality characteristics at UPC site are comparable to the corresponding data monitored at rural or suburban sites of European Mediterranean countries.

Table 6. Summary of statistical air pollution data at UPC and other sites in Greece

STATION	UPC	ALIARTOS[1]	LYKOVRISI	PATISION	PIRAEUS	AG. PARASKEVI
Source		http://www.ypeka.gr				
Area – Station type	Background - Suburban	Background - Rural	Suburban	Traffic - Urban	Traffic - Urban	Background - Suburban

Air pollutant	Statistic	UPC 2012	UPC 2013	ALIARTOS 2012	ALIARTOS 2013	LYKOVRISI 2012	LYKOVRISI 2013	PATISION 2012	PATISION 2013	PIRAEUS 2012	PIRAEUS 2013	AG.PARASKEVI 2012	AG.PARASKEVI 2013
O_3 ($\mu g\ m^{-3}$) – 8-h mean values	Mean	96.1		52	44								
	Median	100.5		50	45								
	Max	176.9		131	120	232	164	108	110	120	115	206	167
	98% values < 120 $\mu g\ m^{-3}$	151.0		110	102	135	129	74	76	92	76	135	133
	Completion (%)	73		99.1	99								
NO ($\mu g\ m^{-3}$) – 1-h mean	Mean	0.5											
	Median	0.1				7	8	58	63	18	22	3	3
	Max	8.1				277	285	586	747	371	405	61	45
	98% values < 120 $\mu g\ m^{-3}$	3.1				105	99	285	303	176	188	7	7
	Completion (%)	84				93	77	94	100	78	72	90	97
NO_2 ($\mu g\ m^{-3}$) – 1-h mean[2]	Mean	5.7		2.74	3.0								
	Median	4.8		2.1	2.4	16	16	61	53	40	34	7	6
	Max	31.0		11.3	22.5	122	113	190	135	121	112	57	57
	98% values <	17.0		7.6	9.7	68	66	127	94	86	80	30	29
	Completion (%)	84		30.9	99.5	93	77	94	100	78	72	90	97
SO_2 ($\mu g\ m^{-3}$) – 24-h mean	Mean			2	1.5			7	7	10	10		
	Median			1.5	1			5	7	9	9		
	Max	N. D.[3]		51.5	59			365	24	32	36		
	98% values <			6	4			22	16	22	24		
	Completion (%)			99.4	99.2			100	100	80	97		
PM_{10} ($\mu g\ m^{-3}$) – 24-h mean	Mean	20.4	14.0										
	Median	18.9	11.5			32	36			38	33	22	23
	Max	91.3	98.7			189	236			90	220	86	193
	98% values <	45.9	53.5			100	116			71	87	41	77
	Completion (%)	70	76			86	98			65	96	79	96
$PM_{2.5}$ ($\mu g\ m^{-3}$) – 24-h	Mean	12.7	10.2										
	Median	12.5	9.2			21	9					14	9
	Max	50.8	42.5			90	39					28	26
	98% values <	27.5	26.1			46	34					24	91
	Completion (%)	70	77			90	50					87	98
CO ($\mu g\ m^{-3}$) – 1-h mean values	Mean			1.9									
	Median			34.8				1200	1100	600	700		
	Max			327.8				9600	10500	6400	8500		
	98% values <			136.3				4200	4700	2600	2900		
	Completion (%)			74.9				100	100	78	98		

[1] Mean Annual Value

[2] The values of UPC refer to 8-h mean concentrations

[3] Not Detectable

46

Table 7. Summary of statistical data (daily) of air pollutants at UPC and other Mediterranean monitoring stations of Europe

		Greece	Malta			Cyprus	Italy					
	Station City	UPC	Zeitun	Msida	Gharb	Limassol	Crotone	Palermo	Palermo	Villasor	Assemini	Locri
	Population (inh.)	28,000	13,133	8,770	1,533	101,000	60,157	655,979	655,979	7,065	26,620	12.459
	Station European Code		MT00004	MT00005	MT00007	CY0005A	IT2031A	IT1691A	IT1076A	IT1307A	IT2049A	IT1940A
	Type of Station	Background	Background	Traffic	Background	Background	Background	Traffic	Background	Background	Traffic	Background
	Type of Area	Suburban	Urban	Urban	Rural	Urban	Urban	Suburban	Suburban	Suburban	Urban	Suburban
	Altitude (m)	61	56	2	114	22	10	69	141	20	5	10
Air pollutant	Year	2012/2013	2012	2012	2012	2011	2011	2011	2011	2011	2011	2011
PM_{10}	Annual Mea-Day ($\mu g\ m^{-3}$)	20.4/14.0	27.7	40.7	25.6	35.7	27.8	25.8	18.0	30.0	26.5	27.9
	98th Percentile ($\mu g\ m^{-3}$)	45.8/53.5	69.1	78.0	59.0	73.5	68	62.0	50.0	65.0	67.0	64.0
	Max ($\mu g\ m^{-3}$)	91.3/98.7	191.1	286.2	216.6	242.3	98.0	87.0	65.0	128.0	71.0	86.0
	Exceedances >50µg m⁻³	4/7	7	54	11	36	27	6	3	31	8	26
	Completion (%)	70/76	34	81	93	91	76	41	45	96	23	90
$PM_{2.5}$	Annual Mean-Day ($\mu g\ m^{-3}$)	12.7/10.2	10.4	18.4	12.7	22.3	18.7					14.2
	98th Percentile ($\mu g\ m^{-3}$)	27.5/26.1	23.1	35.5	28.5	39.6	44.0					27.0
	Max ($\mu g\ m^{-3}$)	50.8/42.5	80.4	48.4	95.9	59.2	113.0					44.0
	Completion (%)	70/74	93	83	74	100	73					93
NO	Daily Mean ($\mu g\ m^{-3}$)	/0.5	2.7	25.1	0.1			8.3	3.1	2.5	6.0	9.9
	98th Percentile ($\mu g\ m^{-3}$)	/1.9	8.7	85.8	0.4			18.4	7.2	13.3	27.8	35.0
	Max ($\mu g\ m^{-3}$)	/3.2	19.6	139.0	1.0			18.4	9.5	20.4	42.3	46.0
	Completion (%)	/84	90	72	96			2	40	95	24	97
NO_2	Daily Mean ($\mu g\ m^{-3}$)	/5.7	22.6	37.1	2.7		34.7	14.6	20.6	12.6	20.7	18.9
	98th Percentile ($\mu g\ m^{-3}$)	/12.0	39.4	74.5	6.7		131.5	28.8	41.6	29.5	49.4	39.8
	Max ($\mu g\ m^{-3}$)	/16.2	50.1	108.6	8.2		205.3	28.8	50.3	33.6	49.6	44.1
	Completion (%)	/84	90	72	96		44	2	40	95	24	97
NO_x	Daily Mean ($\mu g\ m^{-3}$)	/6.2	26.8	75.6	2.9		47.2	21.9	24.7	16.5	29.9	31.7
	98th Percentile ($\mu g\ m^{-3}$)	/13.6	50.2	205.2	7.5		169.7	31.4	50.6	48.3	82.8	83.3
	Max ($\mu g\ m^{-3}$)	/18.0	76.3	274.8	9.2		271.2	31.4	59.0	59.3	114.4	104.1
	Completion (%)	/84	90	72	96		44	2	40	95	24	97

Table 7 (continued)

	Greece	Malta	Malta	Malta	Cyprus	Italy	Italy	Italy	Italy	Italy	Italy
Country	Greece	Malta			Cyprus	Italy					
Station City	UPC	Zeitun	Msida	Gharb	Limassol	Crotone	Palermo	Palermo	Villasor	Assemini	Locri
Population (inh.)	28,000	13,133	8,770	1,533	101,000	60,157	655,979	655,979	7,065	26,620	12.459
Station European Code		MT00004	MT00005	MT00007	CY0005A	IT2031A	IT1691A	IT1076A	IT1307A	IT2049A	IT1940A
Type of Station	Background	Background	Traffic	Background	Background	Background	Traffic	Background	Background	Traffic	Background
Type of Area	Suburban	Urban	Urban	Rural	Urban	Urban	Suburban	Suburban	Suburban	Urban	Suburban
Altitude (m)	61	56	2	114	22	10	69	141	20	5	10
Year	2012/2013	2012	2012	2012	2011	2011	2011	2011	2011	2011	2011
Air pollutant O$_3$											
Annual Mean – Daily 8h max (µg m^{-3})	/116.2	92.0	73.9	105.6					69.6		88.3
98th Percentile (µg m^{-3})	/165.0	134.9	106.0	147.4					92.6		126.0
Max (µg m^{-3})	/176.9	160.1	128.9	163.4					98.9		132.9
Exceedances >120µg m^{-3}	/148	36	1	70					0		18
Completion (%)	/76	90	79	92					24		94
CO											
Annual Mean – Daily 8h max (mg m^{-3})	/0.0	0.4	1.9	0.2		0.9	0.3	0.2			0.7
98th Percentile (mg m^{-3})	/0.1	0.7	2.4	0.3		1.7	0.7	0.3			1.2
Max (mg m^{-3})	/0.4	1.0	3.6	0.3		1.8	1.3	0.8			1.5
Completion (%)	/77	72	73	95		44	27	37			95
SO$_2$											
Daily Mean – Day (µg m^{-3})	3.7	3.0	1.3			9.3	0.6	1.3	0.7	1.3	5.2
98th Percentile (µg m^{-3})	10.2	18.4	4.5			80.0	2.9	4.9	1,5	2.4	14.8
Max (µg m^{-3})	22.1	71.6	6.7			124.4	7.2	6.3	2,9	2.7	15.9
Exceedances >125µg m^{-3}	0	0	0			0	0	0	0		0
Completion (%)	92	85	97			34	46	48	98	24	96

Table 7 (continued)

	Portugal		Spain							France
Country										
Station City	Monte Velho	Malpique	San Jorge	Benidorm	Marbella			Ciutadela de Memorca	Viladecans	Marsellie
Population (inh.)	26,171	4,881		73,768	140,473			29,580	65,188	1,720,941
Station European Code	PT04002	PT05008	ES1542A	ES1675A	ES1657A	ES1750A	ES1393A	ES0006R	ES1903A	FR03043
Type of Station	Background	Background	Background	Background	Background	Background	Traffic	Background	Traffic	Background
Type of Area	Rural	Urban	Rural	Suburban	Suburban	Urban	Urban	Rural	Sub-urban	Urban
Altitude (m)	43	45	181	44	39	36	51	78	14	77
Year	2011	2011	2011	2011	2011	2011	2011	2011	2011	2011
PM_{10} Annual Mean-Day ($\mu g\ m^{-3}$)	21.9	21.3	16.3	10.4	12.8	34.9	28.6	12.9	22.9	29.3
98th Percentile ($\mu g\ m^{-3}$)	44.5	45.5	35.0	25.7	27.0	55.0	52.0	28.0	44.6	55.6
Max ($\mu g\ m^{-3}$)	52.1	51.6	46.0	44.3	30.0	60.0	53.0	33.0	44.9	71.5
Exceedances >50$\mu g\ m^{-3}$	3	1	0	0	0	7	2	0	0	19
Completion (%)	98	41	96	99	100	15	15	95	27	97
$PM_{2.5}$ Annual Mean-Day ($\mu g\ m^{-3}$)	10.6		9.7	8.9	15.2	13.5	16.9		14.1	17.1
98th Percentile ($\mu g\ m^{-3}$)	23.3		21.3	24.4	37.0	34.0	33.0		34.0	38.0
Max ($\mu g\ m^{-3}$)	31.2		26.0	42.8	62.0	49.0	39.0		38.4	49.6
Completion (%)	88		68	99	31	30	48			99.5
NO Daily Mean ($\mu g\ m^{-3}$)			3.0	3.7	4.7	12.2	8.8	0.2	9.5	8.2
98th Percentile ($\mu g\ m^{-3}$)			7.9	9.9	10.0	60.8	35.5	1.4	37.0	43.0
Max ($\mu g\ m^{-3}$)			9.8	15.5	13.8	89.1	49.0	2.7	60.0	73.9
Completion (%)			92	98	98	93	95		94	95
NO_2 Daily Mean ($\mu g\ m^{-3}$)			5.9	8.3	15.5	30.4	28.0	3.9	18.1	32.1
98th Percentile ($\mu g\ m^{-3}$)			11.8	21.0	29.5	65.0	53.3	11.9	41.4	62.4
Max ($\mu g\ m^{-3}$)			15.3	29.4	39.4	80.9	64.1	15.0	60.6	80.1
Completion (%)			92	98	98	93	95	99	92	95
NO_x Daily Mean ($\mu g\ m^{-3}$)			10.5	13.9	22.6		41.5	4.2	32.7	44.7
98th Percentile ($\mu g\ m^{-3}$)			18.6	33.7	44.3		106.9	13.9	85.9	132.0
Max ($\mu g\ m^{-3}$)			20.3	53.2	53.0		139.2	17.8	135.9	192.8
Completion (%)			92	98	98		95	99	92	95

49

Table 7 (continued)

	Portugal		Spain							France
Country										
Station City	Monte Velho	Malpique	San Jorge	Benidorm	Marbella			Ciutadela de Memorca	Viladecans	Marsellie
Population (inh.)	26,171	4,881		73,768	140,473			29,580	65,188	1,720,941
Station European Code	PT04002	PT05008	ES1542A	ES1675A	ES1657A	ES1750A	ES1393A	ES0006R	ES1903A	FR03043
Type of Station	Background	Background	Background	Background	Background	Background	Traffic	Background	Traffic	Background
Type of Area	Rural	Urban	Rural	Suburban	Suburban	Urban	Urban	Rural	Sub-urban	Urban
Air Altitude (m)	43	45	181	44	39	36	51	78	14	77
pollutant Year	2011	2011	2011	2011	2011	2011	2011	2011	2011	2011
O$_3$ Annual Mean – Daily 8h										
max (µg m^{-3})	87.9		95.1	92.0	84.8	82.1	69.4	90.5	81.4	72.7
98th Percentile (µg m^{-3})	122.7		142.0	127.4	122.1	119.3	101.4	123.0	138.1	125.6
Max (µg m^{-3})	139.2		148.6	135.0	133.1	137.4	113.3	131.3	155.8	134.0
Exceedances >120µg m^{-3}	10		47	22	10	7	0	16	123	16
Completion (%)	78		98	98	100	100	94	96	93	100
CO Annual Mean–Daily 8h										
max (mg m^{-3})	0.2			0.1	0.5	0.5	0.4		0.4	0.181[1]
98th Percentile (mg m^{-3})	0.2			0.2	0.7	1.1	0.7		0.9	1.0
Max (mg m^{-3})	0.4			0.3	0.9	1.4	0.8		1.1	1.4
Completion (%)	99			99	100	100	95		97	99
SO$_2$ Daily Mean–Day (µg m^{-3})	3.7	13.5	2.4	2.4	6.4	10.7	8.3	0.3	3.0	1.2
98th Percentile (µg m^{-3})	6.4	16.7	5.3	4.5	9.1	15.6	10.0	0.9	6.0	7.0
Max (µg m^{-3})	7.4	17.9	8.5	10.7	10.3	18.3	10.8	1.6	7.5	14.2
Exceedances >125µg m^{-3}	0	0	0	0	0	0	0	0	0	0
Completion (%)	99	44	88	97	100	100	95	99	87	100

[1] Year 2011

Table 8. Summary of statistical data (annual) of air pollutants at UPC and other Mediterranean monitoring stations of Europe

	Greece	Malta			Cyprus	Italy					
Station City	UPC	Zeitun	Msida	Gharb	Limassol	Crotone	Palermo	Palermo	Villasor	Assemini	Locri
Population (inh.)	28,000	13,133	8,770	1,533	101,000	60,157	655,979	655,979	7,065	26,620	12.459
Station European Code		MT00004	MT00005	MT00007	CY0005A	IT2031A	IT1691A	IT1076A	IT1307A	IT2049A	IT1940A
Type of Station	Background	Background	Traffic	Background	Background	Background	Traffic	Background	Background	Traffic	Background
Type of Area	Suburban	Urban	Urban	Rural	Urban	Urban	Suburban	Suburban	Suburban	Urban	Suburban
Altitude (m)	61	56	2	114	22	10	69	141	20	5	10
Air pollutant Year	2012/2013	2012	2012	2012	2011	2011	2011	2011	2011	2011	2011
PM$_{10}$ Annual Mean-Hour ($\mu g\,m^{-3}$)	20.6/14.1										
98th Percentile ($\mu g\,m^{-3}$)	53.1/52.6										
Max ($\mu g\,m^{-3}$)	386.5/368.1										
Completion (%)	67/74										
PM$_{2.5}$ Annual Mean-Hour ($\mu g\,m^{-3}$)	12.9/10.3										
98th Percentile ($\mu g\,m^{-3}$)	31.1/29.8										
Max ($\mu g\,m^{-3}$)	157/143.9										
Completion (%)	67/74										
NO Hourly Mean ($\mu g\,m^{-3}$)		2.8	25.2	0.1			9.4	3.1	2.5	5.9	10.0
98th Percentile ($\mu g\,m^{-3}$)		20.6	155.3	0.9			20.0	11.0	22.0	45.0	46.0
Max ($\mu g\,m^{-3}$)		117.2	607.0	6.2			24.0	75.0	102.0	228.0	183.0
Completion (%)		92	75	97			2	41	90	23	98
NO$_2$ Hourly Mean ($\mu g\,m^{-3}$)		22.8	37.3	2.7		34.8	15.9	20.4	12.5	20.6	19.0
98th Percentile ($\mu g\,m^{-3}$)		70.9	104.9	6.7		140.0	54.0	71.0	38.0	70.0	50.0
Max ($\mu g\,m^{-3}$)		110.0	199.7	8.2		458.0	77.0	146.0	72.0	124.0	90.0
Exceedances >200µg m^{-3}		0	0	0		30	0	0	0	0	0
Completion (%)		92	75	96		47	2	41	90	23	98
NO$_x$ Hourly Mean ($\mu g\,m^{-3}$)		27.0	76.0	2.9		47.2	24.4	24.6	16.5	29.7	31.9
98th Percentile ($\mu g\,m^{-3}$)		102.4	336.9	10.3		217.0	68.0	82.0	67.0	135.0	109.0
Max ($\mu g\,m^{-3}$)		263.7	1085.5	39.3		674.0	91.0	186.0	191.0	474.0	343.0
Completion (%)		92	75	97		47	2	41	90	23	98

51

Table 8 (continued)

Air pollutant	Greece	Malta			Cyprus	Italy					
Station City	UPC	Zeitun	Msida	Gharb	Limassol	Crotone	Palermo	Palermo	Villasor	Assemini	Locri
Population (inh.)	28,000	13,133	8,770	1,533	101,000	60,157	655,979	655,979	7,065	26,620	12,459
Station European Code		MT00004	MT00005	MT00007	CY0005A	IT2031A	IT1691A	IT1076A	IT1307A	IT2049A	IT1940A
Type of Station	Background	Background	Traffic	Background	Background	Background	Traffic	Background	Background	Traffic	Background
Type of Area	Suburban	Urban	Urban	Rural	Urban	Urban	Suburban	Suburban	Suburban	Urban	Suburban
Altitude (m)	61	56	2	114	22	10	69	141	20	5	10
Year	2012/2013	2012	2012	2012	2011	2011	2011	2011	2011	2011	2011
O_3											
Annual Mean-Hour ($\mu g\,m^{-3}$)	/98.1	75.4	56.8	96.4				96.7		46.6	71.0
98th Percentile ($\mu g\,m^{-3}$)	/151.0	128.1	101.3	138.1				143.0		93.0	121.0
Max ($\mu g\,m^{-3}$)	/176.9	169.8	150.3	177.9				171.0		109.0	149.0
Completion (%)	/73	92	81	93				46		23	94
CO											
Annual Mean-Hour ($mg\,m^{-3}$)	/0.0	0.2	0.5	0.2		0.8	0.3	0.1			0.5
98th Percentile ($mg\,m^{-3}$)	/0.1	0.7	2.3	0.2		2.5	1.4	0.3			1.0
Max ($mg\,m^{-3}$)	/0.3	1.7	6.5	0.6		7.1	2.7	1.4			2.4
Completion (%)	/75	75	76	97		56	29	39			95
SO_2											
Annual Mean-Hour ($\mu g\,m^{-3}$)		3.7	3.0	1.3		12.2	0.8	1.3	0.7	1.3	5.2
98th Percentile ($\mu g\,m^{-3}$)		14.8	16.8	6.0		101.0	5.0	6.0	2.0	4.0	14.8
Max ($\mu g\,m^{-3}$)		92.3	176.3	38.8		929.0	15.0	13.0	14.0	15.0	15.9
Completion (%)		93	87	97		43	46	46	94	23	96

Table 8 (continued)

	Portugal		Spain							France
Station City	Monte Velho	Malpique	San Jorge	Benidorm	Marbella			Ciutadela de Memorca	Viladecans	Marsellie
Population (inh.)	26,171	4,881		73,768	140,473			29,580	65,188	1,720,941
Station European Code	PT04002	PT05008	ES1542A	ES1675A	ES1657A	ES1750A	ES1393A	ES0006R	ES1903A	FR03043
Type of Station	Background	Background	Background	Background	Background	Background	Traffic	Background	Traffic	Babckground
Type of Area	Rural	Urban	Rural	Suburban	Suburban	Urban	Urban	Rural	Suburban	Urban
Altitude (m)	43	45	181	44	39	36	51	78	14	77
Air pollutant Year	2011	2011	2011	2011	2011	2011	2011	2011	2011	2011
PM_{10} Annual Mean-Hour ($\mu g\ m^{-3}$)	22.0	21.2		10.4						29.3
98th Percentile ($\mu g\ m^{-3}$)	58.0	55.5		31.0						72.0
Max ($\mu g\ m^{-3}$)	184.0	73.3		61.0						167.0
Completion (%)	99	41		99						96
$PM_{2.5}$ Annual Mean-Hour ($\mu g\ m^{-3}$)	10.6		9.7	8.9						17.0
98th Percentile ($\mu g\ m^{-3}$)	34.0		37.0	28.0						47.0
Max ($\mu g\ m^{-3}$)	128.0		139.0	59.0						118.0
Completion (%)	90		73	99						99
NO Hourly Mean ($\mu g\ m^{-3}$)			3.0	3.7	4.7	12.2	8.8	0.2	9.4	8.1
98th Percentile ($\mu g\ m^{-3}$)			9.0	16.0	14.0	109.0	58.0	2.1	60.0	73.0
Max ($\mu g\ m^{-3}$)			25.0	147.0	95.0	474.0	223.0	22.4	231.0	355.0
Completion (%)			93	98	97	92	96	98	94	95
NO_2 Hourly Mean ($\mu g\ m^{-3}$)			5.8	8.3	15.5	30.3	28.0	3.9	18.0	32.1
98th Percentile ($\mu g\ m^{-3}$)			15.0	31.0	46.0	97.0	83.0	20.3	60.0	88.0
Max ($\mu g\ m^{-3}$)			29.0	89.0	88.0	180.0	121.0	61.8	110.0	142.0
Exceedances >200$\mu g\ m^{-3}$			0	0	0	0	0	0	0	0
Completion (%)			93	98	97	92	96	98	93	95
NO_x Hourly Mean ($\mu g\ m^{-3}$)			10.5	13.9	22.6		41.5	4.2	32.6	44.6
98th Percentile ($\mu g\ m^{-3}$)			22.1	53.4	64.2		165.6	23.5	142.3	195.7
Max ($\mu g\ m^{-3}$)			62.8	311.4	227.7		453.9	96.2	463.2	676.3
Completion (%)			93	98	97		96	98	93	95

Table 8 (continued)

	Portugal		Spain							France
Country										
Station City	Monte Velho	Malpique	San Jorge	Benidorm	Marbella			Ciutadela de Memorca	Viladecans	Marsellie
Population (inh.)	26,171	4,881		73,768	140,473			29,580	65,188	1,720,941
Station European Code	PT04002	PT05008	ES1542A	ES1675A	ES1657A	ES1750A	ES1393A	ES0006R	ES1903A	FR03043
Type of Station	Background	Background	Background	Background	Background	Background	Traffic	Background	Traffic	Babkground
Type of Area	Rural	Urban	Rural	Suburban	Suburban	Urban	Urban	Rural	Suburban	Urban
Altitude (m)	43	45	181	44	39	36	51	78	14	77
Year	2011	2011	2011	2011	2011	2011	2011	2011	2011	2011
Air pollutant										
O_3 — Annual Mean-Hour ($\mu g\,m^{-3}$)			74.6	79.7	65.4	57.8	52.7	80.8	56.7	51.4
O_3 — 98th Percentile ($\mu g\,m^{-3}$)			129.0	121.0	114.0	115.0	99.0	118.5	127.0	121.0
O_3 — Max ($\mu g\,m^{-3}$)			162.0	153.0	148.0	165.0	124.0	146.5	181.0	177.0
O_3 — Completion (%)			99	98	99	99	95	98	95	100
CO — Annual Mean-Hour ($mg\,m^{-3}$)	0.1			0.1	0.5	0.4	0.3		0.3	0.1
CO — 98th Percentile ($mg\,m^{-3}$)	0.3			0.2	0.7	1.0	0.6		0.8	1.0
CO — Max ($mg\,m^{-3}$)	0.8			0.9	1.0	2.2	1.6		1.5	2.0
CO — Completion (%)	100			99	99	99	96		97	99
SO_2 — Annual Mean-Hour ($\mu g\,m^{-3}$)	3.7	13.6	2.4	2.4	6.4	10.7	8.2	0.3	3.0	1.2
SO_2 — 98th Percentile ($\mu g\,m^{-3}$)	7.0	21.0	7.0	6.0	10.0	18.0	10.0	1.1	7.0	10.0
SO_2 — Max ($\mu g\,m^{-3}$)	32.0	99.8	14.0	59.0	29.0	65.0	15.0	8.8	16.0	123.0
SO_2 — Completion (%)	100	44	91	98	99	96	96	98	89	100

6. CONCLUSIONS

The assessment of air pollution of the University of Patras Campus (UPC) is based on the concentration values of air pollutants measured by the Station of the Environmental Engineering Laboratory (EEL) of the Civil Engineering Department. The values of CO and SO_2 measured were quite low, showing that the UPC site resembles to a rural background site and these air pollutants do not matter the air quality of UPC. Regarding PM_{10} and $PM_{2.5}$, the daily average concentrations exceeded 11 and 13 times, respectively, the enacted limits, although the number of exceedances was within the acceptable limits of legislation (Directive 2008/50/ EC). The levels of PM_1 appear low, but there is no related national or European legislation.

Increased O_3 levels have been monitored, as there have been 148 exceedances of the limit value occurred in an interval less than a year. The affordable number of exceedances was 25 times per calendar year averaged over three years. However, no exceedances of the awareness and alert threshold for O_3 occurred. The levels of NO_x were below the enacted limits and, therefore, they do not matter the air quality of the UPC area.

The O_3 UPC diurnal levels follow well enough the pattern of the background site of Finocalia, Crete. Although the process of formation of the secondary air pollutants in Patras seems to be rather moderate, there exist some extraordinary events of O_3 formation in the afternoon hours. Therefore, UPC site becomes necessary for secondary air pollution observations.

The comparison of the statistical data monitored at the UPC site with corresponding data of other Hellenic sites showed that both the primary and secondary air quality characteristics measured at UPC compare well to the background suburban site of Ag. Paraskevi, Athens. Comparisons of statistical data for the daily or annually concentration values show that both the primary and secondary air quality characteristics at UPC site are comparable to the corresponding data monitored at rural or suburban sites of European Mediterranean countries with approximately same latitude.

ACKNOWLEDGEMENTS

Authors thank Prof. George Angelopoulos, Director of the Materials and Metallurgy Laboratory of the Department of Chemical Engineering of the University of Patras, for his support with the nitrogen oxides analyzer and collaboration in the present program. They also thank Prof. Demetrios Theodoracopoulos, Director of Transportation Works Laboratory, for his financial support.

REFERENCES

Bachmann, J. (2007). "Will the Circle Be Unbroken: A History of the U.S. National Ambient Air Quality Standards", *J. Air & Waste Manage. Assoc.*, 57, pp. 652-697

Danalatos D. and S. Glavas (1996), "Diurnal and Seasonal Variations of Surface Ozone in a Mediterranean Coastal Site, Patras, Greece", *Sci. Tot. Env.* 177, 291-301.

Directive 2008/50/EC of the European Parliament and of the Council of 21 May 2008 on ambient air quality and cleaner air for Europe. *Official Journal of the Europe 11.6.2008,* I.152, pp. 1-44

Gehrig R., Hill M. and Buchmann M. (2004), "Separate determination of PM_{10} emission factors of road traffic for tailpipe emissions and emissions from abrasion and resuspension processes", *Int. J. Environment and Pollution*, 22(3), pp 312-325

Gemmer M. and Xiao B. (2013). "Air Quality Legislation and Standards in the European Union: Background, Status and Public Participation", *Advanced in climate change research* 4(1) pp. 50-59

Gerasopoulos E., Kouvarakis G., Vrekoussis M., Donoussis C., Mihalopoulos N. and Kanakidou M. (2006). "Photochemical ozone production in the Eastern Mediterranean", Atmospheric Environment, Vol. 40, pp 3057-3069

European Environmental Agency (EEA), 2013. *Air Quality in Europe – 2013 Report.* European Union (EU-27) ISBN 978-92-9213-406-8.

Kalabokas, P. D. and Repapis, C. C. (2004). "A climatological study of rural surface ozone in central Greece", *Atmos. Chem. Phys.*, 4, pp. 1139-1147

Ketzel M., Omstedtb G., Johanssonc C., Düringe I., Pohjolaf M., Oettlg D., Gidhagenb L., Wåhlina P., Lohmeyere A., Haakanaf M., Berkowicz R. (2007) "Estimation and validation of $PM_{2.5}/PM_{10}$ exhaust and non-exhaust emission factors for practical street pollution modelling", *Atmospheric Environment*, 41(40), pp. 9370-9385

Saitanis C.J. (2003). "Background ozone monitoring and phytodetection in the greater rural area of Corinth – Greece", *Chemosphere*, 51(9), pp. 913-923

Saitanis, C.J. and Karandinos, M.G. (2001). "Instrumental recording and biomonitoring of ambient ozone in Greek countryside", *Chemosphere*, 44: 813-821

US-EPA, 2009. *Integrated Science Assessment for Particulate Matter*. EPA/600/R-08/139F

World Health Organization (WHO), 2006. *Air Quality Guidelines - Global Update 2005*. ISBN 92 890 2192 6.

World Health Organization (WHO). 2008. *Health risks of ozone from long-range transboundary air pollution*, ISBN 978 92 890 42895

World Health Organization (WHO), 2013a. *Health effects of Particulate Matter*, ISBN 978 92 890 0001 7

World Health Organization (WHO), 2013b. *Review of evidence n health aspects of air pollution – REVIHAAP Project*. WHO Regional Office for Europe, Denmark

Yannopoulos P. C. (2014), "A cost-effective methodology for spatial concentration distributions of urban air pollutants", *Water Air Soil Pollut*, 225(7):1989, DOI: 10.1007/s11270-014-1989-7, pp 1-25.

Web pages:

http://www.ypeka.gr (accessed November 2014, in Greek)

http://www.ploigos.gr (accessed November 2014)

http://www.air-quality.org.uk (accessed November 2014)

http://www.epa.gov (accessed November 2014)

https://earthdata.nasa.gov/labs/worldview/ (accessed November 2014)

http://www.eea.europa.eu/data-and-maps/data/airbase-the-european-air-quality-database-8 (accessed November 2014)